Advances in

INSECT PHYSIOLOGY

VOLUME 6

Advances in

Insect Physiology

Edited by

J. W. L. BEAMENT, J. E. TREHERNE
and V. B. WIGGLESWORTH

Department of Zoology, The University, Cambridge, England

VOLUME 6

1969

ACADEMIC PRESS

London and New York

ACADEMIC PRESS INC. (LONDON) LTD
BERKELEY SQUARE HOUSE
BERKELEY SQUARE
LONDON, W1X 6BA

U.S. Edition published by

ACADEMIC PRESS INC.
111 FIFTH AVENUE
NEW YORK, NEW YORK 10003

Library of Congress Catalog Card Number: 63-14039
SBN: 12–024206–0

Printed in Great Britain by The Whitefriars Press Ltd.,
London and Tonbridge

List of Contributors to Volume 6

E. ASAHINA, *The Institute of Low Temperature Science, Hokkaido University, Sapporo, Japan* (p. 1)

A. D. CARLSON, *Department of Biological Sciences, State University of New York at Stony Brook, Stony Brook, New York, U.S.A.* (p. 51)

JOHN S. EDWARDS, *Department of Zoology, University of Washington, Seattle, Washington, U.S.A.* (p. 97)

RUDOLF HARMSEN, *Biology Department, Queen's University, Kingston, Canada* (p. 139)

P. N. R. USHERWOOD, *Department of Zoology, University of Glasgow, Glasgow, Scotland* (p. 203)

IRMGARD ZIEGLER, *Botanical Institute, Darmstadt Institute of Technology, Darmstadt, Germany* (p. 139)

Acknowledgements

The Editors wish to thank the publishers of *Arch. ital. Biol.*, *Wilhelm Roux Arch.*, *J. Insect Physiol.*, *Comp. Biochem. Physiol.*, *Nature, Lond.*, *Biol. Bull.*, *J. exp. Biol.* and *J. Cell Biol.* for their kind permission to reproduce figures from their publications.

The authors in this volume have been instructed to obtain necessary permissions for use of published figures and the Editors sincerely regret the omission in case any of the sources were overlooked.

J. W. L. Beament
J. E. Treherne
V. B. Wigglesworth

Contents

FROST RESISTANCE IN INSECTS

E. ASAHINA

NEURAL CONTROL OF FIREFLY LUMINESCENCE

A. D. CARLSON

POSTEMBRYONIC DEVELOPMENT AND REGENERATION
OF THE INSECT NERVOUS SYSTEM

JOHN S. EDWARDS

CONTENTS

THE BIOLOGY OF PTERIDINES IN INSECTS
IRMGARD ZIEGLER and RUDOLF HARMSEN

I. Introduction 140
II. Properties and Characteristics of Pteridines 140
III. Separation, Identification and Localization 144
IV. Occurrence in Insects 146
   A. Pigments 146
   B. Colourless Pterines 152
   C. Hydrogenated Pterines 152
V. Localization in the Tissues 160
   A. In the Integument 160
   B. In the Compound Eye 161
   C. In Other Tissues 165
VI. Metabolism of Insect Pterines 165
VII. Hydrogenated Pterines as Co-factors 170
VIII. Relation to Other Pigments 172
   A. Melanin and Sclerotin (= Protein Tanned by Quinones) 172
   B. Ommochromes 173
   C. Flavines 174
   D. Purines 175
IX. Developmental Physiology 175
X. Biosynthesis 177
   A. Biosynthesis of the Pteridine Ring 177
   B. Biosynthesis of C-6 Substituted Pterines 179
   C. Biosynthesis of Other Pterines 182
XI. Physiological Roles of Pterines in the Insect 185
   A. As Co-factors and Growth Substances 185
   B. As Eye Pigments 186
   C. As Metabolic End Products 187
XII. Conclusion and General Discussion 190
References 191

ELECTROCHEMISTRY OF INSECT MUSCLE
P. N. R. USHERWOOD

I. Introduction 205
II. The Muscle Membrane 208
   A. Structure and Morphology 208
   B. Electrical Properties 210
   C. Permeability Properties 211
III. The Extracellular and Intracellular Environments 214
IV. The Resting Membrane Potential 222

Frost Resistance in Insects

E. ASAHINA

*The Institute of Low Temperature Science,
Hokkaido University, Sapporo, Japan*

I. INTRODUCTION

Insects hibernating in temperate and cold regions are, as a rule, able to withstand climatic low temperatures, although they may suffer serious injury if the cold weather is too severe or too prolonged. They can avoid such cold injury by two common means: super-cooling and frost resistance. The former is the avoidance of freezing; the latter the tolerance to freezing. Cold resistance or cold hardiness in overwintering insects is chiefly a matter of the avoidance of freezing (Wigglesworth, 1965, p. 610). In many of the insects ice formation in the insect body is fatal, while others can survive such freezing. The former group is called freezing-susceptible and the latter, freezing-resistant (Salt, 1956) or frost resistant. In a freezing-susceptible insect, cooling to temperatures below the supercooling point, i.e. the highest temperature at which spontaneous freezing occurs in the insect's body, invariably means death, although a frost-resistant insect would survive this event. Such differences in resistance to freezing are usually related to the developmental stages of the insect. In other words, when insects are in the active growth or reproductive stages, they are generally susceptible to freezing. Some of them, however, may become frost resistant, usually with the approach of the cold season. In some insects this frost resistance can also be produced by artificial chilling. Such processes of natural or artificial induction of frost resistance in organisms have been called frost hardening.

Cold resistance in insects has so far been reviewed several times (Uvarov, 1931; Luyet and Gehenio, 1940; Ushatinskaya, 1957; Asahina, 1959b, 1966; Salt, 1961a). In these literatures, however, discussions on the resistance mechanism have often confused the tolerance and avoidance of freezing in insects, since a highly frost-resistant insect is very frequently found to have a high ability to supercool, although the reverse is not true. This chapter will mainly deal with problems of frost tolerance, that is, the resistance to injuries resulting from actual ice formation in the insect body, with emphasis on the evidence from recent work published after 1950. Since Salt has recently presented an elaborate discussion of the supercooling phenomena in insects in his excellent review (1961a, 1966a), the subject of frost avoidance will not be mentioned here except that which is specifically related to the immediate problem.

II. INITIATION OF FREEZING

Supercooling is not necessarily dangerous to many insects. In fact, most hibernating insects can pass the winter in a supercooled state so stable that no mechanical shock will initiate freezing unless the degree of supercooling is considerable (Salt, 1958a). Freezing in an insect's body can be introduced by either (1) spontaneous freezing in some part of the body, or (2) inoculation through the cuticle from an external ice crystal.

A. SPONTANEOUS FREEZING

Spontaneous freezing in very pure water depends entirely upon the chance of an aggregation of water molecules attaining the critical size of a crystal nucleus. In most water phases, freezing involves heterogeneous nucleation in which an ice nucleus is formed by the addition of supercooled water molecules to the surface of a non-aqueous material or *mote* (Dorsey, 1948). The probability of such nucleation increases in every case in which the volume of water increases, the temperature is lowered or the time is lengthened. This is also true of freezing in insects (Salt, 1958a, 1966a). In this respect, it is very interesting to note that, in many overwintering insects, even in those which possess no glycerol or other protective substances, the supercooling point measured during very slow cooling frequently occurs at temperatures a few degrees below $-20°C$ (Aoki and Shinozaki, 1953; Salt, 1953, 1966b, 1966c; Asahina unpublished), which is considered to be the most probable range for inducing nucleation in bulk water in a vessel (Smith-Johansen, 1948; Ôura, 1950). Various tissues in an individual insect may vary in their ability to supercool, but when the tissue which supercools least freezes, the others are inoculated from it and also freeze. In an insect's body, blood in the bulk space of the body cavity may be the most easily freezable fraction.

A very low supercooling point, approaching $-47°C$, has been found in some hibernating insects with high solute concentrations in their blood (Salt, 1956, 1959b). Most larvae or nymphs in the active feeding stage, on the other hand, usually exhibit fairly high supercooling points. This is probably attributable to nucleation in the digestive tract by food and its contaminants (Salt, 1953).

B. INOCULATION BY EXTERNAL ICE CRYSTALS

At subfreezing temperatures, intimate contact between the body wall and an external ice crystal is perhaps the easiest way to initiate freezing in an insect. For this purpose, freezing of the water moistening the surface of the insect's body is most effective (Salt, 1936), while contact between the dry surface of a cooled insect's body and an ice crystal is rather without effect (Aoki and Shinozaki, 1953). In some insects, however, even when the water moistening the body surface has frozen, no ice is formed in the insect's body for a period which may be longer than a few days. Recently Salt suggested that such a very slow inoculation process may be explained by the limitation to ice growth in the minute pathways through the body wall of the insect (Salt, 1963). The problem of ice seeding in a whole insect may be one of the most interesting topics from an ecological viewpoint.

Except in aquatic insects, freezing in insects in natural surroundings is generally presumed to be induced by spontaneous freezing, since effective external ice seeding of even naked insects is rare. Furthermore, the various coverings found in overwintering insects such as hibernacula, cocoons, puparia and egg shells can serve as effective barriers to ice seeding, although they are generally ineffective in influencing the rate of cooling in the insects, at least under outdoor conditions.

III. THE FREEZING PROCESS

A. FREEZING OF THE BLOOD

The process of the development of ice crystals in insect blood has been described in detail as it occurs in the prepupa of the slug caterpillar, *Monema flavescens* (Asahina, 1953b). There is no apparent difference between ice crystal forms in pure water and ice caterpillar blood at a high subfreezing temperature. However, crystal growth in the blood, especially in the overwintering period, is much slower than in pure water (Shinozaki, 1954a).

With inoculation, the ice originates as a small, discoid shape and develops into hexagonal, fern-like crystals (Fig. 1A). At temperatures very near the freezing point, however, inoculation with an ice-tipped pipette is unsuccessful in the blood of an overwintering insect. The ice grows only on the tip of the pipette by withdrawing water from the drop of blood, and when the pipette is removed all of the ice is withdrawn with it. In the blood of larvae in the summer stage, when they are susceptible to freezing, dendritic ice branches, once formed, rapidly fuse with each other, embedding the blood in the ice mass in many vein-like branches or droplets (Fig. 1B). In the blood of prepupa in the winter, when the insect is frost resistant, the ice grows slowly and the concentrated blood is not usually embedded throughout the ice since the branches of the ice crystals do not easily fuse into large ice masses (Fig. 1C). In such cases, dark areas surrounding the ice crystals, attributable to the concentration of blood, are usually observed to remain after rapid thawing of the ice. If the formation of such layers of concentrated blood play a role in the frost resistance of the insect, this advantageous situation is most effectively achieved in the insect in the overwintering stage (see Section V, A).

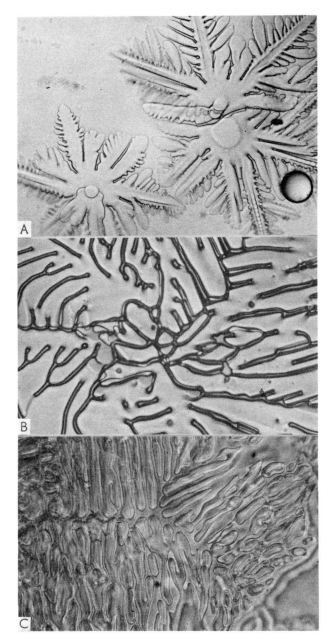

Fig. 1. Ice formation in the blood of the prepupa of *M. flavescens* (from Asahina, 1953b).

(A)   Blood of the overwintering insect, 15 seconds after ice seeding at −2.6°C × 106.
(B)   Blood of the summer insect, 3 days after spinning the cocoon, frozen at −8°C × 106.
(C)   Blood of the overwintering insect, frozen spontaneously at −20°C × 375.

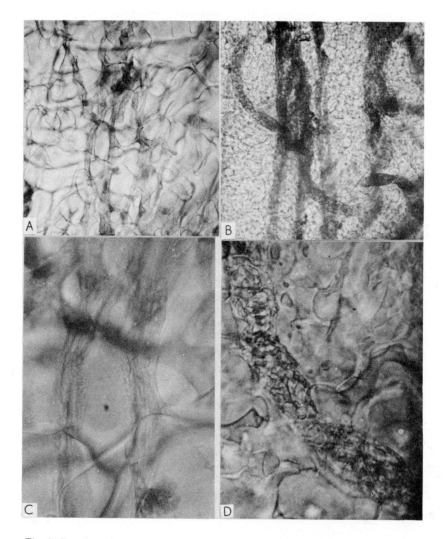

Fig. 2. Freezing of pericardial cells with alary muscle in the overwintering prepupa of *M. flavescens.*

(A)   Extracellularly frozen cells at $-4.2°C$ × 106.

(B)   Intracellularly frozen cells at $-10°C$ × 106.

(C)   Detail of (A) showing the unfrozen interior of the cell cooled to $-7°C$ × 360.

(D)   Detail of (B) showing formation of large ice masses due to the recrystallization within the cell, 11 min after intracellular freezing at $-10°C$ × 360 (A, B and D, from Asahina *et al.,* 1954; C, from Asahina, 1959b).

B. TYPES OF CELL FREEZING

A variety of insect cells have been observed under the microscope during freezing (Asahina *et al.,* 1954; Salt, 1959a, 1961b; Asahina, unpublished). Two types of cell freezing, extra- and intracellular, were clearly observed. The former, differing only in degree, occurs in every case of freezing, while the latter, if it occurs at all, takes place subsequent to extracellular freezing, except in cases of very rapidly cooled cells. This suggests that intracellular freezing is induced by inoculation from the outside of the cell with ice crystals at moderately low temperatures (Asahina, 1961).

With cooling, ice usually forms first in the blood surrounding the cells with or without external seeding. As freezing proceeds, the ice develops around the cell which suffers dehydration and contraction (Fig. 2A, C). In many of the tissue cells from various insects, such extracellularly frozen cells can regain their normal appearance and activity following thawing unless the freezing has been at too low a temperature or too prolonged. The degree of resistance to extracellular freezing is various according to the kind of tissue cells, even within an individual insect. Cells from frost-resistant insects can very frequently endure a very long period of freezing at fairly low temperatures. On the other hand, tissues from freezing-susceptible insects are very liable to intracellular freezing, when the entire cell darkens almost instantaneously shortly after it is covered with extracellular ice. Cells so frozen never contract (Fig. 2B, D). Such an instantaneous blackening is the result of the very rapid formation throughout the cell of numerous minute ice crystals. This has been called "flashing" (Luyet and Gibbs, 1937; Asahina, 1953, 1956). The fine ice crystals formed by "flashing" are very apt to fuse subsequently into larger ice masses (Fig. 2D), especially at high subfreezing temperatures.

The likelihood of intracellular ice formation depends upon factors which control the rate of removal of water from the cell in the early process of extracelular freezing, involving cooling rate, surface-volume ratio, membrane permeability to water, and the temperature coefficient of the permeability constant (Asahina, 1961, 1962; Mazur, 1963). An easier occurrence of intracellular freezing is, therefore, more reasonably expected in large spherical cells, such as fat body cells, than in many kinds of small slender form cells.

When intracellularly frozen cells are thawed, there is evidence of destruction of the cell structure. Vacuolization or swelling of the protoplasm is common in cells thawed following a very short period of freezing, while those thawed after freezing for a few hours or more are frequently observed to have a network of coagulated cytoplasm. A change in the colour of vitally stained cells is invariably found in both cases. These changes in the protoplasmic pattern of cells frozen intracellularly always occur, regardless of the existence of protective substances in the insects involved (see Section V, B, 2).

Intracellular freezing has been generally regarded as fatal in various living cells at climatic low temperatures (Chambers and Hale, 1932; Asahina et al., 1954; Asahina, 1956; Meryman, 1956). In insect cells, however, some recent reports have given contrary evidence (Salt, 1959a, 1961b; Losina-Losinsky, 1963b). Survival of intracellularly frozen and thawed fat body cells in the larvae of a gallfly, *Eurosta solidaginis,* was first reported by Salt (1959a). It is difficult, however, to ascertain whether or not the frozen and thawed fat cell is actually alive because of the large number of oil droplets which fill the entire cell. Moreover, the greater frost injury in rapidly frozen insects than in slowly frozen ones (Salt, 1961b) can be more easily explained by the increase in the number of intracellularly frozen cells in the former than by any other interpretations. A result recently obtained by Tanno (1968a) from a freezing experiment with poplar sawflies strongly suggests that even in fat body cells intracellular freezing is fatal (see Section IV, B). It seems, however, probable that intracellular freezing may occur in at least some small fraction of the tissue cells of some insects which are easily supercooled to certain low temperatures particularly when artificial cooling is applied.

C. POSSIBLE PROCESS OF FREEZING IN AN INTACT INSECT

The freezing curve of an insect may provide a useful clue to the freezing process of the entire animal. The freezing curve of the individual insect differs very little whether freezing is spontaneous or by inoculation with an external ice crystal, though the degree of supercooling is usually far smaller in the latter. The typical freezing curve of a relatively large insect is presented in Fig. 6. On this curve, a rapid fall in the insect's body temperature to the so-called supercooling point is followed by a sharp rise which results from the onset of internal freezing. This sharp rise and the plateau which follows it appears to result primarily from ice formation in the blood filling the body cavity, since water in bulk space is presumed to freeze easily and rapidly. This was confirmed by changes in the shapes of the freezing curves of insects in which solutions other than blood had been injected into the body cavity (Shinozaki, 1954b). When the cooling is rapid or the insect is small, only a rebound or a peak appears instead of the plateau shown in Fig. 6.

As the amount of ice increases in the body cavity, the blood is concentrated. The tissue cells become dehydrated as extracellular freezing proceeds in the insect. When almost all the blood in the body cavity has been frozen, extracellular freezing still continues between the tissue cells but the rate of ice formation rapidly decreases. This decrease results in a gradual fall in the freezing curve. As the temperature continues to fall, one or more rebounds are sometimes observed (Payne, 1927; Asahina, 1959b) which may be caused by a small fraction of body tissue which has remained in a supercooled state.

It may be safely said that freezing of a slowly cooled whole insect, even when it follows supercooling of some ten degrees or more, is generally of the extracellular type (Asahina et al., 1954; Salt, 1962; Tanno, 1968a). In tissue cells, especially those from frost-resistant insects, which are cooled under the microscope in a hanging drop of blood, intracellular freezing is very rarely observed unless cooling is very rapid. Intracellular freezing can be observed only when freezing of the preparation is induced at a temperature far lower, usually 10°C or more, than the freezing point of the blood, if the amount of blood which is surrounding the cells is very small. Besides, freezing curves in a variety of insects have proved that even in very tiny insects such as silkworm eggs, a considerable rise in body temperature, sometimes estimated to exceed 10°C, occurs with spontaneous freezing when the cooling is slow (Aoki, 1962; Asahina and Tanno, 1967). Frost-resistant insects, as a rule, show no abnormalities in the tissue cells when they have revived from freezing, with the exception of the fat body cells in which small oil droplets of uniform size frequently fuse into larger homogeneous masses.

The observations strongly suggest that in a whole insect, at least in a frost-resistant one, the prevention of intracellular freezing in tissue cells is greatly favoured by the freezing of a large amount of blood prior to cell freezing. This is also supported by recent histological studies of frozen insects by the use of freeze-sectioning (Asahina and Tanno, 1967; Tanno, 1968a). Frost-resistant prepupae of the poplar sawfly, Trichiocampus populi, in which blood has its freezing point at about −2°C, were cooled slowly in air at −20°C, when spontaneous freezing occurred in the insect at about −8°C. Cross sections of the frozen insects invariably showed that even very large fat body cells froze extracellularly (Tanno, 1968a) (Fig. 3). Besides, in some tissues in these insects, the individual tissue as a whole was observed to be dehydrated and contracted as seen in an extracellularly frozen cell (Tanno, 1968a).

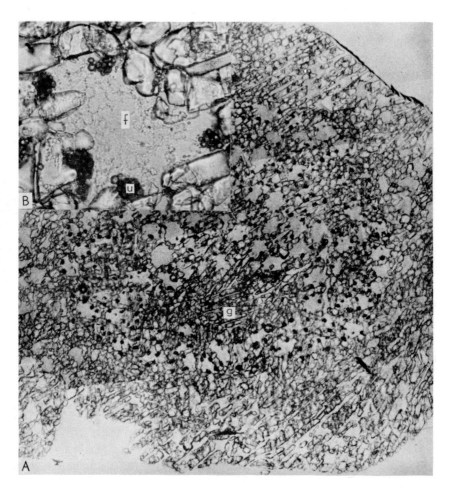

Fig. 3.

(A)   Cross-section of the frozen prepupa of the sawfly, *T. populi*, slowly cooled at a
        rate of 0.4°C/min g, Gut. x 37.
(B)   Detail of (A) showing an extracellularly frozen fat body cell (f) in the visceral
        layer surrounded by large ice particles. Dark masses (u) indicate pouches filled
        with spherulites of uric acid in the surface layer of the fat body cell. x 235.
        (Modified from Tanno, 1968.)

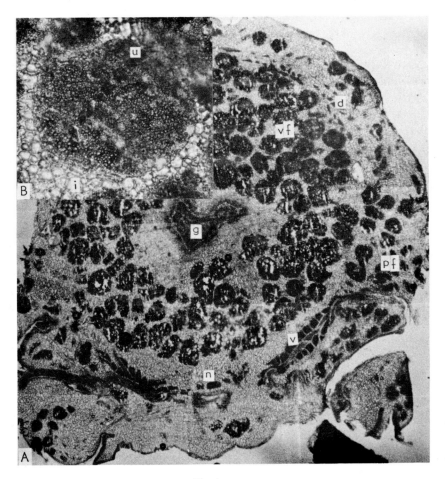

Fig. 4.

(A)   Cross-section of the frozen prepupa of *T. populi,* very rapidly cooled at a rate of 327°C/min, showing intracellular ice formation (darkening) in all of the tissue cells. d, Dorsal longitudinal muscle; g, Gut; n, Ventral nerve cord; pf, Fat body cells in the parietal layer; v, Ventral longitudinal muscle; vf, Fat body cells in the visceral layer. x 37.

(B)   Detail of (A) showing an intracellularly frozen fat body cell surrounded by many small ice particles (i). u, A pouch filled with uric acid spherulites. x 235.
(Modified from Tanno, 1968a.)

D. AMOUNT OF ICE FORMED IN AN INSECT

A quantitative determination of ice formed in a freezing insect at successively lower temperatures may be a very useful method of analysing the freezing process of the insect, since the amount of ice increases as the cooling proceeds. Many insects, however, are very apt to supercool to fairly low temperatures. This sometimes leads to a misunderstanding of the amount of ice in an insect at a given freezing temperature. In this respect, Sacharov's report (1939) that the frozen bodies of some frost-resistant insects still contain a considerable amount of unfreezable water at a temperature as low as $-17°C$ is open to question, since the dilatometric technique he used is by no means suitable for insects. Using a calorimetric method, Ditman *et al.* (1942) found no difference in either the percentage of total water unfreezable, or the unfreezable water–dry weight ratio between freezing-susceptible and freezing-resistant forms when they were frozen to temperatures below their supercooling points. In this work, however, ice determination was not made in individual materials, but in groups of insects. This, unfortunately, led them to the assumption that body freezing in an insect, at any temperature lower than the supercooling point, always caused the maximum freezing of body water. However, all of their results can be safely explained on the basis that the amount of ice formed in an insect is a function of the temperature. This is supported by all of the recent works in which the amount of ice in a frozen insect has been investigated (Scholander *et al.,* 1953; Salt, 1955; Shinozaki, 1962). The amount of unfrozen water decreases with decreasing temperature in a hyperbolic manner until it becomes nearly constant at a temperature around $-30°C$ (Fig. 5). Since the freezing point of the blood is the major factor determining the amount of ice formed in the insect (Shinozaki, 1962), it is suggested that low molecular weight solutions dominate and that the amount of bound water is scanty in insects (Salt, 1955).

Recently Shinozaki (1962) determined in detail the amount of ice formed in the prepupa of a slug caterpillar, *M. flavescens*, throughout a period from autumn to the following summer. He found that there was a clear seasonal variation in the amount of ice formed in the insect (Fig. 17). Such variations can easily be explained by changes in the freezing point of the blood, resulting from fluctuation in the concentration of glycerol produced by the insect (see Section V, B, 2).

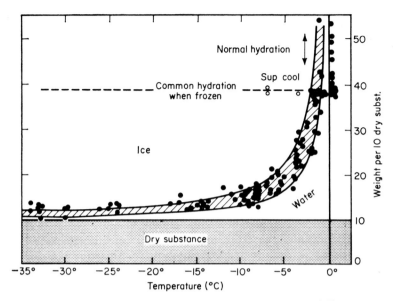

Fig. 5. The relation of unfrozen water and ice to the dry weight of *Chironomus* larvae at graded freezing temperatures (from Scholander *et al.*, 1953).

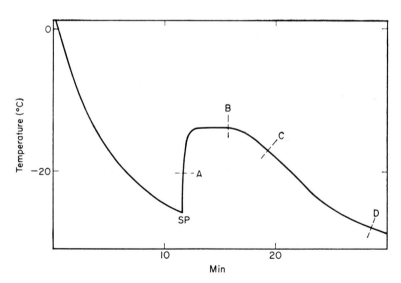

Fig. 6. Freezing curve of an overwintering pupa of *Papilio xuthus*. SP, Supercooling point. See text for explanation of points A, B, C and D (partly modified from Tanno, 1963).

IV. FROST INJURY

A. PROCESS OF INJURY

When cooling is so rapid that a large proportion of the tissue cells of the freezing insect may be expected to freeze intracellularly, the insect always dies immediately after thawing, usually with obvious destruction of some of the tissue cells. However, even in slowly cooled insects in which extracellular freezing may be generally expected to occur, frost injury increases as freezing proceeds in the insect. By rapidly thawing an insect at a certain stage in the process of freezing, Bachmetjew (1901) first tried to find the critical temperature on the freezing curve at which the freezing insect is fatally injured. Although his work has been criticized by many workers (Uvarov, 1931), the method employed seems to be very useful in investigating the process of frost injury in individual insects. In various freezing-susceptible insects, body freezing up to around point A on the freezing curve presented in Fig. 6 has been found to be nearly harmless (Aoki, 1956, 1962; Tanno, 1963). Even freezing to point B on the same curve is never fatal and the insects resume development after thawing, though some degree of frost injury, such as a failure of moulting and of formation of normal adult tissues frequently occurs. Frozen insects thawed at point C usually suffer serious injury, although some of them can survive and sometimes develop to abnormal imagoes (Tanno, 1963). Freezing of susceptible insects to point D invariably results in immediate frost-killing. These observations suggest that frost injury in insects increasingly takes place as freezing of tissue cells proceeds, even though it may be extracellular.

Frost-resistant or frost-hardy insects, on the other hand, can survive freezing down to point D on the freezing curve or to even lower temperatures. The lower limit of temperature which they can survive varies with the degree of frost resistance of the insect used. The very hardy prepupa of the moth, *M. flavescens,* can withstand severe freezing in various liquid gases; the moderately hardy larva of the chafer, *Cetonia roetotsi,* can tolerate freezing temperatures to $-20°C$; and the slightly hardy overwintering caterpillars of various kinds of cutworms can survive freezing only at temperatures above $-10°C$, at least for one day (see Table III, p. 28). However, if the freezing is too severe or too prolonged even these insects suffer the same kind of frost injuries described above in freezing-susceptible insects.

B. NATURE OF INJURY

Frost killing of an insect does not always take place immediately after thawing. Active heart beats are frequently observed after thawing, although the insect may die within several days or more. In the actively feeding summer stage, a brown coloration of the blood is commonly observed in frozen and thawed insects, even when they are still alive with an active heart beat. This suggests the oxidation of tyrosine in the blood possibly resulting from the destruction of some tissue cells by freezing. These insects with brown blood generally die within a few days after thawing. The most common sublethal frost injury, as mentioned before, is the failure to complete metamorphosis. Many overwintering larvae or pupae are frequently found to be apparently normal for many days after thawing and even feed, yet often cannot develop further and may remain in the larval or pupal stage sometimes for a few years (Asahina and Takehara, 1964). Some can transform into adult insects under their pupal cuticles but are unable to shed the cuticles successfully. Half imago formation is the most interesting example of such frost injuries. This is frequently observed in the overwintering pupae of some Papilionid butterflies of both freezing-susceptible and freezing-resistant forms. After relatively severe freezing, the thawed pupae apparently develop normally. In some of these insects, however, the formation of adult tissue is restricted to the anterior half of the pupal bodies; the posterior half, behind the third or fourth abdominal segment, remains in the pupal state and survives for some 10 days, even after the anterior half has died (Fig. 7). Even in the remarkably frost-resistant insect, *Papilio machaon,* most of the pupae which survive freezing in liquid oxygen have been found to become half imagoes (Asahina, 1959a).

Regarding the nature of freezing injury in a whole insect, recently Tanno (1967a, 1968a) made very interesting observations on the Japanese poplar sawfly, *T. populi.* The overwintering prepupae of this insect is remarkably frost resistant; it can survive freezing even in liquid nitrogen, as long as the cooling is very slow (Asahina and Tanno, 1964). Clear morphological differences were observed in the prepupae between the fat body cells in the visceral layer and those in the parietal layer: both are nearly spherical in shape; however, the former is much larger in volume, with an average diameter of 230 μ, than the latter, which has a diameter of about 170 μ (Tanno, 1965). Occurrence of intracellular freezing is, therefore, easier in the former than in the latter, although both of them are rather apt to freeze internally when compared with other tissue cells in the insect. It was

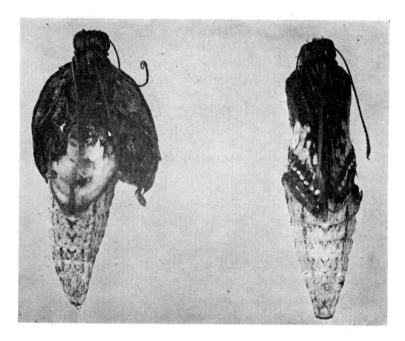

Fig. 7. Half-imagoes of *Papilio machaon* emerged from pupae immersed in liquid oxygen for two days after pre-freezing at −30°C. The inside of unmetamorphosed abdomen is seen in the insect on the right x 3 (from Asahina, 1966).

also confirmed in the prepupae that the fat body cells in the parietal layer (parietal f. b. cells) were mostly consumed during the prepupal stage prior to the transformation to pupa, while those in the visceral layer (Visceral f. b. cells) were mainly consumed throughout the metamorphosis from prepupa to imago (Tanno, 1967a). When the insect was frozen at an appropriate cooling rate (4°C/min), nearly all the visceral f. b. cells froze intracellularly, while only less than one fifth of the total number of the parietal f. b. cells froze intracellularly. The prepupa frozen in this manner did not transform to pupa and died within some fifty days after thawing. A slower rate of cooling (0.8°C/min) in the freezing prepupa caused intracellular freezing in about one sixth of the total number of the visceral f. b. cells, but in none of the parietal f. b. cells. This resulted in the successful adult formation under their pupal cuticles in many frozen-thawed prepupae, but about half of the number failed to moult. When they were frozen at a cooling rate (0.4°C/min) slow enough to cause no intracellular freezing in any fat body cells (Fig. 3), all of the prepupae successfully completed normal metamorphosis after thawing. On the other hand, if they were frozen very rapidly (327°C/min) and intracellular freezing was observed to occur in all fat body cells (Fig. 4), all of the prepupae were found to be dead immediately after thawing with an apparent destruction of fat body cells in the insects (Tanno, 1968a). From the observations described above, it seems likely that in the larval or pupal stage an occurrence of freezing damage in some fraction of fat body cells, which are the most susceptible parts in the frozen insects, results in a failure to complete metamorphosis.

Clearly, frost resistance in a whole insect is limited by the frost-resistance capacity of those tissue cells most susceptible to freezing. Muscular tissues, especially heart with alary muscle are, as a rule, very resistant to freezing even in insects in the summer stage. The isolated heart can very frequently resume rhythmical contraction after freeze-thawing at an apparently lower temperature than that tolerated by the whole insect from which the heart was isolated. In adult insects bodily freezing almost invariably results in fatal injury, even at high subzero temperatures, although some of them are frequently found to be apparently normal immediately after thawing. This presumably reflects the relative inability of some well differentiated tissues in an adult insect to resist frost injury.

One popular hypothesis of frost injury in animal tissue is the so-called "salt injury theory" originally developed in relation to mammalian erythrocytes (Lovelock, 1953a). This hypothesis proposes that the dominant lethal factor during freezing is the increasing concentration of electrolytes both inside and outside the cell. However, this seems unlikely in the case of insects. Isolated hearts from insects in both summer and winter have been proved to survive freezing in salt solutions with a high concentration of sodium chloride (Asahina et al., 1954; Wilbur and McMahan, 1958). In many insects the concentration of electrolytes in the blood is fairly high (Duchâteau et al., 1953; Stobbart and Shaw, 1964), although not as high as in mammals. Many insects are frost resistant even though they lack any appreciable amount of "protective substance", an innocuous small molecular compound which acts as a salt buffer (Takehara and Asahina, 1960b; Sømme, 1964) (Table III).

Injury to living cells freezing in a salt solution has been known to increase rapidly within a certain temperature range, usually much higher than the eutectic point (Smith, 1961). But in many moderately frost-resistant insects, with or without a "protective substance", the lower the freezing temperature, the greater the injury. An adverse temperature range around $-50°C$, has recently been reported for the remarkably frost-resistant larvae of the European corn borer, *Pyrausta nubilalis* (Losina-Losinsky, 1962). However, in the slug caterpillar, *M. flavescens,* freezing experiments at temperatures down to $-70°C$ have not disclosed any such critical range where injuries rapidly develop (Asahina and Takehara, 1964). In addition, slow rewarming from $-70°C$ or liquid gas temperatures was found to be highly effective in producing a good survival rate in various freezing insects (Asahina and Aoki, 1958b; Asahina and Takehara, 1964; Tanno, 1968b). It has also been demonstrated that in non-frost-resistant insects, an injection of glycerol, a well-known "protective substance", does not increase the insect's capacity to resist body freezing (see Section V, B, 2). One may only assume that severe dehydration of living cells is a primary cause of frost injury in extracellular frozen insects. Such dehydration may possibly cause an unfavourable change in an interaction between protoplasmic protein and surrounding water to maintain a well organized structure of a living protoplasmic system. The mechanism of such injury is not yet clear, although a very interesting hypothesis involving the structural change in the protoplasmic protein during freezing has recently been presented (Levitt, 1962).

C. INJURY RESULTING FROM EXTENDED FREEZING PERIODS

It has been well known that frost-resistant insects can be kept alive in the frozen state for long periods (Losina-Losinsky, 1937; Asahina, 1955; Barnes and Hodson, 1956; Sømme, 1964). Factors affecting these long periods of survival may conceivably involve the metabolism in the frozen insects. Kozhantshikov attempted to determine the respiration in a variety of insects while frozen and reported the discovery of "thermostable respiration" in cold-hardy insects (Kozhantshikov, 1938). However, because of the inadequate procedure used in measuring the oxygen consumption of these frozen insects, and because of the confusion created by using the term "cold-hardy" for both frost avoidance and frost tolerance in insects, his hypothesis has been subject to criticism by recent workers. Scholander and his collaborators (1953) determined the respiration of a remarkably frost-resistant larva of an Alaskan chironomid midge at graded temperatures down to $-15°C$. A precipitous drop in the oxygen consumption of individual larvae was recorded when freezing occurred in the insect; the Q_{10}/O_2 in the temperature range just above the freezing point differed greatly from those following freezing (Fig. 8). Salt (1958b) determined the respiration rate in the larvae

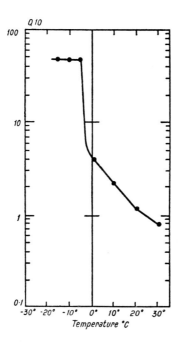

Fig. 8. Change in the respiratory rate of Alaskan chironomid larvae at temperatures between +30 and −15°C, expressed as Q_{10}/O_2 (from Scholander et al., 1953).

of the Mediterranean flour moth, *Anagasta kühniella*, under supercooled conditions. The temperature-oxygen consumption curve he obtained was found to form an extension of the above-zero portion of the curve. The Q_{10}/O_2 in a frozen insect is, therefore, apparently far higher than that in a supercooled one. Such a high Q_{10}/O_2 value is presumably caused by an inhibition of metabolism in the frozen insect, which results not only from the low temperature itself, but from the high degree of dehydration of body tissues and also the hindrance to diffusion created by the ice formed within the insect (Scholander et al., 1953). This seems to be supported by Kanwisher's unpublished results in diapausing pupae of *Hyalophora cecropia*. Even while frozen the Q_{10}/O_2 in these insects was found to be lower than in Scholander's chironomid. This may reasonably be explained by the fact that the diapausing *cecropia* has a large amount of glycerol (Wyatt and Meyer, 1959), resulting in a smaller amount of water being removed by freezing and, consequently, lesser inhibition of the metabolism due to dehydration (Kanwisher, unpublished). In this insect, of course, the oxygen consumption was far lower in the frozen pupa than in the supercooled (Fig. 9).

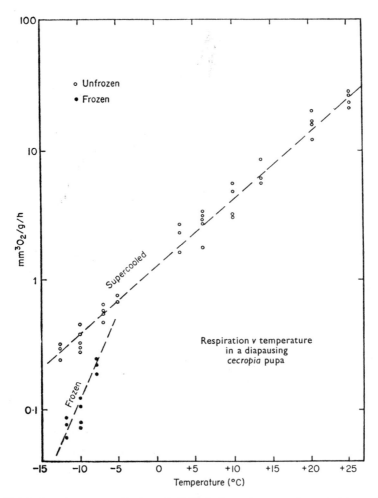

Fig. 9. Oxygen consumption in an individual diapausing *cecropia* pupa at temperatures between +30 and 13°C (by courtesy of Dr. Kanwisher).

From these observations one may assume a potentiality for extra-ordinarily long survival times in frozen insects even at moderately cold temperatures (Scholander *et al.,* 1953). Preliminary studies of the low temperature storage of the overwintering prepupae of *Monema flavescens* may present a clue to the solution of this problem (Table I). These prepupae could be frozen for 100 days at −20° and at −10°C without harmful effect. Longer storage at these

Table I

Effect of subzero temperatures on the life span of frozen or supercooled prepupae of the slug caterpillar, *Monema flavescens.* (Asahina, 1955)

Treatment	Duration of storage (days)	Number of insects used	Survival	No. pupated	No. emerged
Freezing at −10°C	100	10	10	9	9
	150	10	7	6	3
	180	10	5	2	0
	200	10	3	1	0
	225	10	3	0	—
	250	10	1	0	—
Freezing at −20°C	100	10	10	10	8
	150	10	7	7	3
	180	10	5	3	2
	200	10	5	1	1
	225	10	5	1	1
	250	10	3	3	3
Supercooling at −10°C	100	10	10	10	10
	150	10	8	5	3
	180	10	6	1	1
	225	10	7	0	—
	275	10	1	0	—
	350	10	1	0	—
	400	10	0	—	—
Unfrozen control at about 0°C	150	10	10	10	9
	200	10	10	10	9
	275	10	10	10	9

The longest survival time recorded was 374 days in the frozen prepupae stored at −20°C, and more than 435 days in the unfrozen ones kept at 0°C.

temperatures, however, caused injury although storage at −20°C seemed to be less harmful than storage at −10°C (Asahina, 1955). This suggests that injuries which result from prolonged freezing may involve some metabolic process still functioning at these subfreezing temperatures. Contrary to expectation, storage in the frozen state was found to be as harmful as storage in the supercooled state, at least at −10°C. A similar result was also obtained in the diapausing larvae of the European corn borer by Hanec and Beck. They concluded that the degree of supercooling may not be an adequate criterion for the estimation of ability of the insect to survive prolonged exposure to subfreezing conditions (Hanec and Beck, 1960). It is of interest to note that both supercooling and freezing at moderately subzero temperatures were found to be more injurious over long periods of storage than control storage at 0°C, which was entirely harmless throughout a 275-day period (Table I).

From the results presented in Table I it appears that death following a long period of freezing in frost-resistant insects does not result from the exhaustion of energy sources available at ordinary temperatures, but by some metabolic imbalance which occurs below a certain subfreezing temperature. Nevertheless, it seems still possible to preserve frost-resistant insects for very long periods provided they are kept at extremely low temperatures. Unfortunately data on the long-term storage of frozen insects at liquid gas temperatures are still very incomplete. Overwintering prepupae of *M. flavescens* were kept in liquid nitrogen with previous treatment at −30°C (see Section VI). Practically no insects were killed within about a 100 day storage at approximately −190°C, but they failed to complete metamorphosis after thawing. During one year of storage, the rate of survival of the frozen insects gradually decreased, but many survived for several months after thawing, although none of them could achieve metamorphosis. Even after 810 days of storage in liquid nitrogen, in some of the prepupae heart beats could be resumed after warming; they died, however, within a few months (Asahina unpublished).

D. EFFECT OF FREEZING ON THE TERMINATION OF DIAPAUSE

It is well known that cold treatment is very effective in terminating diapause in various insects. This may be attributable to some metabolic processes, called diapause development (Andrewartha, 1952), which occur during the period of chilling (Wigglesworth, 1964). In some insects such diapause development effectively proceeds at temperatures below the freezing points of the body fluid (see

Wigglesworth, 1965, p. 111). However, this presumably occurs in supercooled insects, but not in frozen ones. In the egg of the wheat bulb fly *Leptohylemyia coarctata,* a low temperature of $-24°C$ is highly effective for the completion of diapause. But when the eggs were supercooled to $-26.5°C$ and subsequently frozen, they were observed to be dead after thawing (Way, 1960).

On the other hand, a striking effect of freezing on the termination of diapause was very recently observed by Tanno (1967) in overwintering prepupae of *T. populi.* At the beginning of December the prepupal diapause in the insect was so stable that even chilling at $2°C$ for as long as 4 months failed to induce the resumption of pupal development after warming, but bodily freezing at temperatures below $-15°C$ always resulted in an apparent termination of diapause (Table II).

Table II

Freezing temperatures and the termination of diapause in the overwintering poplar sawfly (Tanno, 1967b)

Freezing for (hrs)	Temperatures ($°C$)						
	Unfrozen 2	-5	-10	-15	-20	-25	-30
0.5	0	0	0	2	6	7	8
24	0	0	0	3	7	7	7

Number of insects developed to pupa at $20°C$ in ten prepupae within 40 days after freeze-thaw treatment is indicated.

In the case of the prepupae overwintering outdoors, they usually underwent freezing and thawing several times during the cold season, since the supercooling point of this insect is as high as $-8.6 \pm 0.4°C$ (Tanno and Asahina, 1964). However, in the winter of 1966–1967 no insect was released from diapause until a cold night in January, when the air temperature first dropped to $-18°C$. Since that night practically all prepupae began to resume metamorphosis when they were transferred to $20°C$ (Tanno, 1967b).

A.I.P. 2

It is very interesting to note that a body freezing for only half an hour is highly effective to terminate the diapause in the sawfly, since cold treatment without freezing has so far been known to become effective only after several hours, or usually several days, of the treatment (Wigglesworth, 1964). Moreover, freezing at temperatures above $-10°C$ never causes termination of diapause in the insect (Table II). These results suggest that the mechanisms involved in the termination of diapause by freezing of the insect may conceivably be different from those by chilling without freezing. It was assumed that only some special organs, such as the corpus allatum, which might be responsible for keeping the insect in diapause (Fukaya and Mitsuhashi, 1961), were very active in a diapausing insect, and therefore the cells in these organs might not survive freezing below $-15°C$, since the cells in active stage were shown to be much more susceptible to freezing than resting cells in various organisms (Tanno, 1967b).

V. MECHANISMS OF FROST RESISTANCE

A. AVOIDANCE OF INTRACELLULAR FREEZING

As previously observed, avoidance of intracellular freezing in most tissue cells seems to be necessary to preserve an insect alive in the frozen state at least at climatic low temperatures. It has been demonstrated in various living cells that, even at high subfreezing temperatures, contact between ice and the cell surface can cause intracellular freezing only when the cells are cooled rapidly (Asahina, 1953a, 1956). This may be explained by an easy freezing of water within the well supercooled protoplasmic layer of the cell surface which otherwise can serve as an excellent barrier to prevent the inoculation of the cell by external ice (Chambers and Hale, 1932; Asahina, 1961). It seems therefore reasonable that every factor which reduces the rate of cooling at the surface of cells will be effective in the prevention of intracellular freezing. In insects, the freezing of a large amount of body fluid prior to cell freezing is undoubtedly beneficial. In this respect, it is of interest to note that almost all of the frost-resistant insects so far known are in the larval or pupal stage (Table III) and in these two stages, as a rule, they have much more blood than in any other stage of their development.

Other important characteristics of tissue cells in frost-resistant organisms may be an increase in their permeability to water and in their ability to alter their form, though neither of these characteristics has yet been studied in insect cells. It has been demonstrated in the egg cells of the sea urchin that these two factors, by increasing the velocity of dehydration from the freezing cell, can reduce the cooling rate at the cell surface by feeding the growing extracellular ice and maintaining the liberation of latent heat (Asahina, 1962).

The blood also seems to have some characteristics which tend to prevent the seeding of ice into the tissue cells. As previously stated (Section III, A), the formation of a concentrated blood layer around the ice crystals was clearly demonstrated in the freezing of insect blood, particularly that from frost-resistant insects (Asahina, 1953b). Such a layer may not only retard the propagation of freezing, but may effectively prevent contact between the growing ice crystals in the body cavity and the cell surfaces (Kistler, 1936). An increase in certain small molecule hydrophilic substances in the blood, a frequent adjunct to frost hardening, lessens the proportion of water frozen at a given low temperature and also facilitates the formation of a thick layer of concentrated blood around the ice crystals in a freezing insect.

B. RESISTANCE TO EXTRACELLULAR FREEZING

After an extracellularly frozen cell has reached a steady state without intracellular ice formation, if the temperature is too low or the exposure too prolonged, the cell protoplasm is liable to injury. Since such injury is presumed to result from dehydration both inside and outside the cells, the following factors have been suggested as responsible for protecting insects against frost injury from extracellular freezing.

1. Quality and Quantity of Water in the Insect

It has long been known that the water content is usually less in a cold-hardy insect than in a non-cold-hardy one (Payne, 1927; Uvarov, 1931). In this respect, Robinson (1928) has proposed a "bound water" theory. He stated that during the process of cold-hardening in insects, the conversion of free water to bound water was much easier in cold-resistant forms. His theory has been severely criticized, mainly because of the erroneous assumption that all free water in a freezing insect is completely frozen at $-20°C$ (Ditman et al., 1942; Asahina, 1959b; Salt, 1961a; Shinozaki, 1962). Even from his own data, supercooling can be found in many insects at $-20°C$.

Salt (1961a) revealed that prior artificial dehydration failed to produce better survival rates in frozen and thawed insects, although less ice was formed. Nevertheless, a good inverse correlation between water content and frost resistance is frequently found in insects. As has been pointed out by Salt (1961a), this may be attributable to combinations of the following sets of circumstances. (1) Frost resistance is usually greater in hibernating stages than in others because during active stages of either growth or reproduction physiological conditions are unfavourable for the development of frost resistance. (2) Water content is usually lower in hibernating stages because of gut-evacuation, the cessation or decrease of various secretory activities, and the accumulation of non-aqueous reserve materials, mainly in the fat body, with consequent reduction of the need for water in other tissues. Shinozaki's recent studies (1962), on the amount of ice formed in slug caterpillars, also demonstrate an apparent decrease in the freezable water at usual climatic low temperatures during the interval from autumn to winter, when the insect becomes remarkably frost resistant. This, however, is apparently not caused by an increase in bound water, but mainly by an increase in the glycerol produced by the insect (see Section V, B, 2).

2. Protective Substances

The discovery of specialized conversion of glycogen to polyhydric alcohols in diapausing insects (Chino, 1958; Wyatt and Meyer, 1959) provided a new basis for the theory of frost resistance. Since the existence of a sufficient amount of small molecule hydrophilic compounds lessens the proportion of water frozen in the whole system, the production of these compounds in overwintering insects has recently been proposed to be responsible for frost resistance (Salt, 1957, 1961a; Smith, 1961). Glycerol has been found in many insects known to be highly frost resistant (Table III), the most remarkable example possibly being the overwintering larva of a parasitic wasp, *Bracon cephi*. An extraordinarily high concentration of glycerol, 25% of the fresh body weight, was found in this insect by Salt (1958c). Only two kinds of polyhydric alcohols (polyols), glycerol and sorbitol, and a sugar, trehalose, are so far known to be protective substances in overwintering insects.

a. Polyols. Chino (1957) found that as eggs of the silkworm, *Bombyx mori,* entered diapause, glycogen was converted to polyols and was later resynthesized from the polyols shortly after the termination of diapause. Similar changes in glycerol content with the stages of diapause were also observed with overwintering eggs of an aphid, *Pterocomma smithia,* and of moths, *Alsophila pometaria* and *Acrolita naevana* (Sømme, 1964, 1965). None of these diapausing eggs with polyol were, however, proved to be frost resistant (Aoki, 1962; Sømme, 1964, 1965).

The relationship of glycerol in insects to frost resistance was studied in detail in the slug caterpillar, *M. flavescens,* with special reference to the effect on glycerol formation of environmental temperature (Takehara and Asahina, 1960a, 1961; Takehara, 1963, 1966). Under outdoor conditions glycerol is formed from glycogen during the autumn and remains at an almost constant level throughout the winter. However, when the insect is warmed in the following spring, glycerol is again completely converted to glycogen (Fig. 10). The total sugar content, on the other hand, rapidly decreases to a

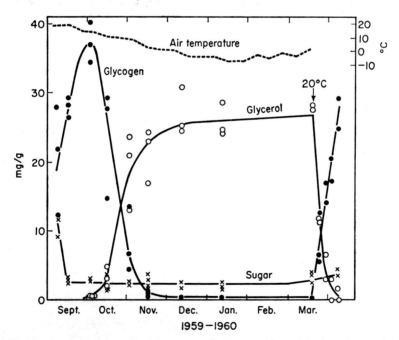

Fig. 10. Seasonal changes in the content of glycogen, glycerol and total sugar in the overwintering prepupae of *M. flavescens.* Arrow denotes the time of transference of the insects to 20°C from outdoor temperatures (from Takehara and Asahina, 1960).

Table III

Frost resistance and glycerol content in insects

Species and stage	Glycerol Approx. maximum per cent of fresh wt.	Freezing Temperature tolerated for 24 hours	Reference
Hemiptera			
Pterocomma smithia, eggs	15.5*	None	Sømme, 1964
Lepidoptera			
Loxostege sticticalis, larvae	4.3	None	Salt, 1957
Laspeyresia strobilella, larvae	17.3*	Lower than −40°C	Sømme, 1965
Pyrausta nubilalis, larvae	4.0	Lower than −40°C	Takehara and Asahina, 1960b
Monema flavescens, prepupae	5.0	Lower than −40°C	Takehara and Asahina, 1961
Alsophila pometaria, eggs	15.1*	None	Sømme, 1964
Philudoria albomaculata, larvae	0.5	−20°C	Asahina, 1966
Agrotis fucosa, larvae	0	− 5°C	Takehara and Asahina, 1959
Amathes ditrapezium, larvae	0.1	−15°C	Ohyama, unpublished
Spilosoma niveus, larvae	1.8	−25°C	Asahina, 1966
Hyalophora cecropia, pupae	3.4	Lower than −70°C	Asahina and Tanno, 1966
Bombyx mori, eggs	1.1**	None	Aoki, 1962
Papilio xuthus, pupae	3.8	None	Tanno, 1963
Papilio machaon, pupae	3.4	Lower than −40°C	Asahina, 1966
Hestina japonica, larvae	0	−15°C	Takehara and Asahina, 1960b
Araschnia levana, pupae	4.8	None	Asahina, unpublished

Coleoptera			
Cetonia roetotsi, larvae	3.8	−20°C	Asahina, 1959b
Dendroctonus monticolae, larvae	23.4*	None	Sømme, 1964
A cerambycid, larvae	2.5	−15°C	Asahina, unpublished
Hymenoptera			
Camponotus obscuripes, adults	5.3	None	Tanno, 1962
Camponotus herculeanus, adults	5.8*	None	Sømme, 1964
Hoplismenus obscurus, adults	3.0	None	Asahina, 1959b
Pterocormus molitorius, adults	4.5	−10°C	Asahina and Tanno, 1968
Bracon cephi, larvae	25.0	Lower than −40°C	Salt, 1959b
Eurytoma gigantea, larvae	23.4*	Lower than −40°C	Sømme, 1964
Megachile rotundata, larvae	2.2*	None	Sømme, 1964
Diptera			
Eurosta solidaginis, larvae	2.0***	Lower than −40°C	Salt, 1957
Euura nodus, larvae	0	Lower than −7°C	Sømme, 1964
Diplolepis sp., prepupae	17.6*	None	Sømme, 1964
Polypedilum vanderplanki, larvae	0	−12°C	Leader, 1962

* Per cent of the sum of the water content plus glycerol.
** Plus 2.2% sorbitol.   *** Plus 4.2% sorbitol.

very small amount in early autumn and remains so throughout the winter. Temperature was found to be one of the most important factors in the control of glycerol metabolism in this insect (Takehara, 1966). The optimum temperature for the rapid production of glycerol in *M. flavescens* appears to be around 10°C. A small amount of glycerol is sometimes produced by a few individuals at 0°C. However, after glycerol has been accumulated to some extent by a previous treatment at 10°C, a storage of the insect at 0°C always results in a very gradual decrease in the glycerol content. Practically no glycerol is formed in the insect reared at 20°C from the time of cocooning, and even when glycerol is artificially injected into such prepupae it completely disappears within about 60 days after injection (Takehara, 1963), although at the same temperature diapause usually persists for more than one year in this insect (Asahina, 1959c). At a constant temperature of 10°C, the glycerol content, after it has reached a maximum, remains at a constant level for about 2 months and then suddenly decreases. The same insects reared outdoors show a remarkable increase in glycerol during autumn, retaining this high glycerol content throughout the 5 months of the cold season when the mean atmospheric temperature continuously remains below 10°C (Fig. 11). Incubation of these insects at 10°C and at 20°C always

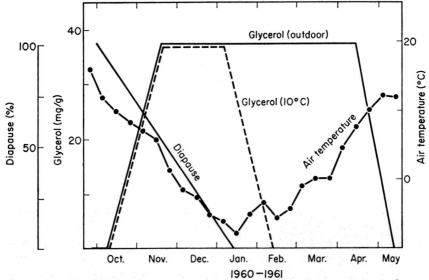

Fig. 11. Changes in glycerol content in the prepupae of *M. flavescens* reared at 10°C and outdoor temperatures, showing the fluctuation of outdoor air temperature and the percentage diapause during the overwintering period (from Takehara and Asahina, 1961).

results in a rapid decrease in their glycerol content (Takehara and Asahina, 1961). Also in many diapausing insects, the rate of increase in glycerol content varies with the temperature to which they are exposed (Chino, 1958; Sømme, 1964; Asahina and Tanno, unpublished) (Fig. 12). In the pupae of the *cecropia* silkworm which have

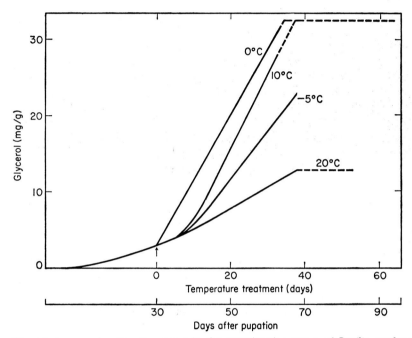

Fig. 12. Increase in glycerol content in the overwintering pupae of *Papilio machaon* reared at various constant temperatures. Arrow indicates time when temperature treatment was initiated. Before this time all the pupae were kept at 20°C (from Asahina, 1966).

been known to produce glycerol at 25°C, a rearing temperature of 6°C was remarkably effective to increase glycerol (Wyatt, 1967).

In many insects glycerol formation has been found to occur only during diapause, but in the adult carpenter ant fluctuation of glycerol content is entirely temperature dependent, even in the actively feeding summer season (Dubach *et al.*, 1959; Tanno, 1962; Sømme, 1964). Chino (1958) stated that in diapausing eggs of the silkworm, the resynthesis of glycogen from polyols was directly associated with the termination of diapause *per se* and not with the subsequent process of post-diapause development. Sømme's recent work (1964) in various overwintering insects seems to support this view, at least in the egg stage. In the larval or pupal stage, however, most glycerolated insects appear to lose glycerol rapidly only when post-diapause development begins. Overwintering prepupae of *M. flavescens,* under outdoor conditions, for example, are completely released from diapause in early winter, usually before the middle of January, but they retain high levels of both glycerol and frost resistance for several months at least, unless they are warmed to temperatures near 10°C (Fig. 11). At a rearing temperature of about 0°C no post-diapause development is found in these insects even for periods as long as 9 months (Asahina, 1955).

On the other hand, current knowledge of glycerolated insects does not always favour the view that the possession of an appreciable amount of glycerol, or its equivalent, is a factor indispensable to frost resistance in insects. It is apparent from Table III that there is no clear correlation between the glycerol content and the degree of frost resistance among different species. Some insects can survive freezing at moderately low temperatures without any glycerol while many cannot survive even with a large amount. To determine the direct effect of glycerol on frost resistance, glycerol was injected into freezing-susceptible forms of diapausing insects which either did or did not possess natural glycerol. The injected glycerol failed to convert a freezing-susceptible insect to a freezing-resistant one, even though it had no harmful effect upon further development in the unfrozen control (Takehara and Asahina, 1960b; Tanno, 1963).

Changes in the degree of frost resistance during periods of both increase and decrease in glycerol content were also observed in the prepupae of *M. flavescens* (Takehara and Asahina, 1961; Asahina and Takehara, 1964). Prior to glycerol formation, the prepupae in cocoons can survive freezing at −10°C for about 1 day. As glycerol increases, their frost resistance is rapidly enhanced. After the glycerol content has reached about 2% of the fresh body weight, further increase appears unnecessary for the insects to maintain their high frost-resistant capacity, since at that point they have become sufficiently resistant to survive freezing, even at liquid gas temperatures (Fig. 13). The pupae of the *cecropia* silkworm, reared at 25°C, can also survive freezing at −20° and −30°C, provided they have a glycerol content of more than 10 mg/ml blood (Asahina and Tanno, 1966). Since a slight increase in glycerol content is always followed by a remarkable increase in frost-resistant capacity, the mechanism of the protective action of glycerol against frost injury in insects may well differ from

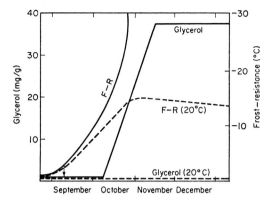

Fig. 13. Increase in frost resistance in the prepupae of *M. flavescens* in autumn, with or without concomitant increase in glycerol. Solid lines denote insects under outdoor conditions. Broken lines denote insects reared at 20°C. Arrow shows the time of cocooning (from Asahina, 1966).

the mechanism operating in the protection of mammalian erythrocytes by glycerol (Lovelock, 1953b). As the glycerol disappears from the prepupae of *M. flavescens,* there is a better correlation between glycerol content and the degree of frost resistance even at a rearing temperature of 10°C, when post-diapause development proceeds very slowly (Fig. 14). It is of interest to note that, even after the prepupae

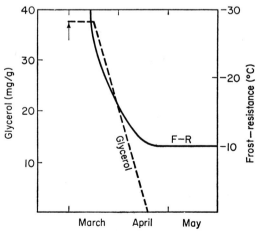

Fig. 14. Decrease in frost resistance during the period when glycerol disappears in the prepupae of *M. flavescens* in spring transferred to 10°C from outdoor temperature conditions. Arrow denotes the time of the transference (from Asahina, 1966).

have no more glycerol, they can survive freezing at $-10°C$ unless post-diapause development proceeds appreciably (Takehara and Asahina, 1961).

In the case of adult insects, only an Ichneumonid, *Pterocormus molitorius,* has so far been proved to be frost resistant (Table III). The glycerol content of this wasp was determined to be 3.2–4.5% of fresh body weight in midwinter (Asahina and Tanno, 1968).

Isolated insect tissues, as in some of the mammalian tissues, can be remarkably protected from freezing injury by artificial glycerolation. The glycerolated heart isolated from an adult Passalid beetle, *Popilius disjunctus,* resumed beating after more than 10 days freezing treatment at $-20°C$ (Wilbur and McMahan, 1958). A similar result was also obtained in the summer season with the isolated hearts from various cutworms (Asahina, unpublished). It must be noted in both these cases, however, that the isolated tissues were frozen with an artificial balanced salt solution, insect Ringer, and that no cutworms injected with glycerol were observed to become frost resistant (Asahina, unpublished). From these results it seems apparent that the possession of small molecule polyols in insects is not the sole factor in their frost resistance, although an increase in these polyols may reasonably enhance the resistance of some insects.

b. Sugars. Small molecule compounds other than polyol, which have been observed to accumulate in overwintering insects, are sugars of mono- or disaccharide. Since these sugars were well-known protective substances against freezing injury in both higher plants (Sakai, 1962) and mammalian blood cells (Doebbler *et al.,* 1966), the quantitative increase of these sugars in insects may possibly favour survival at low temperatures. Fairly high total sugar levels are sometimes found in adult insects (Wyatt, 1967). None of them, however, have been observed to be frost resistant. Certain adult solitary bees, *Ceratina flavipes* and *C. japonica,* contain extraordinarily high sugar levels in winter: 10–15% of fresh body weight, comprising 45–47 mg/g fructose, 30–48 mg/g glucose and about 4 mg/g trehalose (Tanno, 1964). Even these bees could not survive freezing, but the high sugar content appears to contribute to their ability to supercool (Table IV).

A remarkable example of a frost-resistant insect with a high sugar level was recently reported (Tanno and Asahina, 1964; Asahina and Tanno, 1964; Tanno, 1967a). Diapausing prepupa of the Japanese poplar sawfly, *T. populi* accumulates sugars at the beginning of autumn; the total sugar levels reach about 5–7% of fresh body weight,

Table IV

Freezing temperatures of overwintering adult *Ceratina* bees

Order of treatment	Freezing temperature (°C)	Number of frozen insects within 24 hours at a given low temperature*			
		♂		♀	
1	−10	6	30%	7	21%
2	−15	6	30	8	23
3	−23	3	15	10	30
4	−30	5	25	9	26
Total number of insects used		20	100	34	100

*All of the insects were exposed to the first treatment for 24 hours, and the insects remaining unfrozen were transferred to the next.

of which more than 90% was determined to be trehalose. This persists throughout the six months of the cold season, although all of the insects terminate the prepupal diapause in January (Tanno and Asahina, 1964). A clear correlation between sugar content and the frost resistance is observed in the insects during the period from August to June in the following summer. The prepupae have such a resistance during the six months of the cold season that none of them are killed, even in liquid nitrogen, provided they were previously frozen at −30°C (Tanno, 1965c) (see Section VI).

c. Protective Substances and Supercooling Ability. In the majority of overwintering insects the accumulated small molecule compounds, regardless of whether they are polyol or sugar, contribute to their cold resistance by lowering their supercooling points rather than by increasing their resistant ability to freezing, since they usually pass the cold season in a supercooled state. Salt (1958c, 1959b) observed a clear parallelism between glycerol content and the supercooling point in overwintering larvae of the sawfly parasite wasp, *B. cephi.* Similar results were also obtained in several overwintering insects (Sømme, 1964). In most cases, the supercooling point depressions were larger than the expected values from the melting point depressions of the haemolymph, resulting from the increase in glycerol concentrations. A remarkable seasonal change in supercooling point with the change in glycerol content was observed in various insects (Sømme, 1965) (Fig. 15).

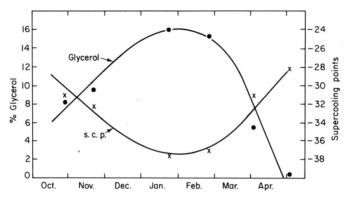

Fig. 15. Changes in glycerol content and supercooling points in larvae of *Laspeyresia strobilella* stored outdoors during the winter (from Sømme, 1965).

3. Protoplasmic Factor

Scholander's Alaskan midge offers striking evidence of remarkable frost resistance in an actively feeding insect (Scholander *et al.*, 1953). This may be an unusual case because an appreciable increase in protective substances in an insect in the feeding or growing stage is presumed difficult, or impossible. This may also be true in the case of a tropical midge from Equatorial Africa, *Polypedilum vanderplanki*, a remarkably drought-resistant insect found by Hinton (1960). The larvae of this midge were found to be frost-resistant even in the actively feeding stage (Leader, 1962), although they possessed no glycerol (Leader, unpublished). Under normal conditions the larvae are only moderately frost-resistant; when once frozen at temperatures below −30°C, they can survive after thawing for only a few days (Leader, 1962). However, if they are partially dehydrated for a few hours by a previous freezing at −10°C, or by immersing them for 4 hours in 1.17 M sucrose solution, some of them can survive freezing at −40°C (Leader, 1962).

It has commonly been assumed that diapause is causally related to cold resistance in general. Based on this assumption, Ushatinskaya (1957) presented a vast amount of data regarding seasonal changes in the components of various diapausing insects. Unfortunately she did not deal with any polyols, which might explain, at least to some extent, the increase in cold resistance of both frost-avoiding and frost-tolerant types of the insects used. However, it seems likely that changes other than small molecule substance formation must occur in diapausing insects to increase their frost-resistant ability since, with or without such a substance, most known frost-resistant larvae or pupae are in the diapause stage, and they rapidly lose this ability when an appreciable post-diapause morphogenesis has taken place. In fact, in the prepupae of *M. flavescens,* an apparent increase in frost resistance was found at the beginning of winter, even in insects reared at the constant temperature of 20°C where neither polyols nor sugars were appreciably produced (Fig. 13). Also, only in those prepupae which have been hardened at 20°C is an injection of glycerol found to be particularly effective in increasing their frost resistance (Asahina and Takehara, 1964). From the above observations it appears likely that other unknown factors are primarily responsible for frost resistance in insects, and that without these factors the mere presence of a protective substance will not cause frost resistance.

Levitt (1962) has recently expounded a new hypothesis of frost injury and resistance in plants based on the assumption of a structural change in protoplasmic protein. A somewhat similar structural change in protoplasm has also been suggested by Asahina and Tanno (1963a, b) in the fertilized egg cell of the sea urchin, in which a remarkably rapid increase in frost resistance was demonstrated without the production of protective small molecule substances. At present these theories are only hypotheses and there are still many unanswered questions. However, one of the most probable causes of increased resistance to fatal dehydration may be some changes in the structure of protoplasm. Furthermore, such a protoplasmic change may contribute to the prevention of intracellular freezing, perhaps through an increase in permeability to water and the ability of the cell surface to deform (see Section V, A). At any rate, the nature of the interaction between water and protoplasmic protein is surely vital in any consideration of the mechanism of frost resistance in living matter, although Robinson's classic bound water theory (1928) of the winter hardiness of insects is certainly untenable (see Section V, B, 1).

C. PROCESS OF FROST HARDENING

As indicated above, the known frost-resistant insects are found exclusively in the larval or pupal stage. They can probably be divided into two types with regard to the nature of frost hardening. One type is represented by the chironomid larvae from both tropical and Arctic regions, the other includes all other insects mentioned here.

The former type of insect can presumably tolerate freezing through-out the larval stage regardless of the season, even when actively feeding, while the latter is freezing-resistant only in the diapausing and post-diapausing stages (see Section V, B, 3) which usually coincide with the overwintering period.

In a diapausing insect the following tentative interpretation of the process of frost hardening can be considered. As the process of diapause proceeds in the insect the tissue cells undergo a change, very possibly in protoplasmic structure, in which the cells become changeable in shape, more permeable to water and also resistant to the dehydration of extracellular freezing. Such a protoplasmic change may also create circumstances in which a large molecule reserve substance such as glycogen may be converted into some small molecule substances with a subsequent increase in protective action.

It has been generally accepted that exposure to cold for relatively long periods of time, usually at a temperature slightly higher than 0°C, is necessary to produce frost resistance in insects (Barnes and Hodson, 1956). Recent work, however, reveals that a fairly high degree of frost resistance can be brought about, at least in some insects, without any cold treatment. Pupae of *cecropia* silkworm (Asahina and Tanno, 1966) and prepupae of the slug caterpillar (Asahina and Takehara, 1964), both in the diapause stage, can apparently be hardened by rearing them at 25°C and 20°C respectively. Pupae of *cecropia* silkworm reared at a constant temperature of 25°C were found to survive freezing at −70°C and subsequently appeared on the wing quite normally (Asahina and Tanno, 1966) (Fig. 16). A reasonable interpretation of the effectiveness of mild chilling in producing good frost hardening in insects may, therefore, be that the optimal temperature range for the rapid formation of protective substance in an insect, frequently, though not always, involves cold temperatures above zero. In addition, keeping insects at cold temperatures aids them in retaining the frost resistance by retarding both the post-diapause development and the conversion of protective substance into a large molecular one.

VI. FREEZING RESISTANCE IN INSECTS AT ARTIFICIAL VERY LOW TEMPERATURES

The lowest climatic temperature is known to be more than −60°C, but is usually between −20° and −50°C in most of the cold regions with insect habitation. Some very frost-resistant insects, however, can survive at far lower temperatures in the laboratory. It is well known that various excised tissues often survive slow freezing even to super-low temperatures (Smith, 1954, p. 18). This may be explained by the innocuous extracellular freezing which occurs in these organisms. Extracellular freezing is easily produced in living cells in wet tissue with slow cooling and intracellular freezing is generally believed to be rare under such conditions. If the freezing injury in highly frost-resistant organisms is mainly caused by intracellular ice formation, they may become resistant to rapid cooling, even at super-low temperatures, following an initial extracellular freezing sufficient to render their cells too dry to freeze internally. To examine this proposition, slow freezing should be applied to the organism prior to rapid cooling to liquid gas temperatures. This method of pre-freezing, or stepwise freezing, is not new. In 1951, De Coninck applied two-step freezing to a free living nematode, *Anguillula silusiae*. He showed the remarkable effect of pre-freezing at $-30°C$ in making the animal tolerant to freezing at $-196°C$. Luyet and Keane (1955) also demonstrated that pre-freezing at temperatures lower than $-27°C$ was distinctly effective in keeping bovine spermatozoa alive when rapidly cooled to $-195°C$. They assumed that the initial freezing of the spermatozoa at the higher temperature definitely afforded protection against freezing injury at lower temperatures, although they did not propose that intracellular freezing might be the lethal element in rapid cooling. Many studies along this line have recently been conducted. Sakai (1956) successfully applied the pre-freezing method to hardy wood twigs. In animal material, this was examined in various intact insects as well as excised tissues or cells (Asahina and Aoki, 1958a; Asahina, 1958, 1959a; Asahina and Tanno, 1966; Tanno, 1968).

The two-step freezing method was found effective for a fairly large chrysalis as well as a tiny larva of a few mg in body weight, provided both were sufficiently resistant to body freezing at an initial temperature of around $-30°C$. Overwintering pupae of the swallow tail, *Papilio machaon,* for example, were kept in a frozen state at $-30°C$ for 1 h and then immersed directly into liquid oxygen for 2 days. Of the ten pupae examined, five were able to resume their development after rewarming, although they were found to become half-imagoes (Fig. 7). In the prepupae of the slug caterpillar, *M. flavescens,* nearly all of those initially frozen at $-30°C$ survived freezing in liquid oxygen and could metamorphose to adult moths under their pupal cuticles although they failed to shed them (Asahina

Fig. 16. *Cecropia* silkworms successfully metamorphosed to adult moths after freeze-thawing at and from very low temperatures during their pupal stage (from Asahina and Tanno, 1966).
(A) Moth from a pupa frozen at -70°C after pre-freezing at -30°C. x 0.5.
(B) Moth from a glycerol-injected pupa frozen very slowly in liquid nitrogen after two-step pre-freezing at -30° and -70°C. The pupal cuticle was artificially removed. x 1.25.

and Aoki, 1958b). Overwintering larvae of the European corn borer pre-frozen at -30°C had a good survival rate after freezing in liquid nitrogen. Even though they possessed an amount of glycerol equivalent to the amount in the slug caterpillar, recovery of the European corn borer from freezing at super-low temperatures was subnormal, and, as a rule, insufficient to allow them to resume development; no larvae could pupate although some survived for as long as 70 days after thawing (Takehara and Asahina, 1960b). This was also the case when the same insect was re-examined by Losina-Losinsky (1962). Besides this method, he tried a three-step freezing in which the larvae were first frozen at -30°C for 1 h, then at -79°C for 1 h and finally transferred into liquid nitrogen. The result, however, was very similar to the mentioned results obtained by two-step freezing (Losina-Losinsky, 1962). Losina-Losinsky also reported the ability to keep the larvae alive in liquid helium with a five-step cooling, although the recovery was very incomplete (Losina-Losinsky, 1963a). In all three of these insect species, the control insects frozen to -30°C for 1 day and thawed without being taken to lower temperatures, suffered no injury (Table V).

Complete recovery of insects cooled to liquid helium temperature was achieved by Hinton (1960) in the remarkably drought-resistant

Table V

Effect of prefreezing temperatures on survival of frozen and thawed insects at and from liquid oxygen temperature (Asahina, 1966).

Prefreezing temperature (°C)	Number of insects used in each case	Number of Surviving insects	
		Pupa of *Papilio machaon*	Prepupa of *Monema flavescens*
−20	10	0 (10)***	0 (9)
−25	10	6* —	2**(10)
−30	10	7* (10)	9**(10)
−40	10	10* —	9**(9)
−50	10	9* (9)	— —

* All surviving insects resumed development but emergence was abnormal, or they became half-imagoes.

** All surviving insects resumed development but were unable to shed pupal skins except for those pre-frozen at −25°C, all of which died within several days after thawing.

*** Number in parenthesis denotes survival of control insects thawed after pre-freezing only at indicated temperatures. Almost all these control insects developed normally after rewarming.

larva of a tropical Chironomid, *Polypedilum vanderplanki.* He gradually desiccated the larva to a water content of about 8% and then transferred it into liquid helium. After rewarming and rehydration, the larva developed into a normal adult (Hinton, 1960).

The highest temperature at which the pre-freezing treatment enables an insect to withstand super-low temperatures seems to be around −25°C (Table V). Nearly all slug caterpillars pre-frozen at either −30° or −40°C withstood rapid cooling in liquid oxygen, while those insects pre-frozen at −25°C died within several days of rewarming, although shortly after thawing some of them were alive with active heart beats (Asahina, 1959a). In the prepupae of the poplar sawfly, which have a large amount of trehalose instead of polyol, pre-freezing even at −20°C was found to be as effective as −30°C (Tanno and Asahina, 1964).

Various insects to which the pre-freezing method had been applied were dissected after rewarming from super-low temperatures to examine any changes which might have taken place in their tissue cells (Asahina and Aoki, 1958b; Asahina, unpublished). Immediately after thawing most of the cells appeared quite normal in those insects which survived, except for the fat body cells in which small oil droplets of uniform size had occasionally fused into larger masses. In the dead insects, however, there usually was a remarkable destruction of the fat body cells, and numerous oil droplets from the destroyed cells were observed floating on the blood surface. In these insects, some of the tissue cells were probably intracellularly frozen, whereas in the insects which survived most of the freezing must have been extracellular. It is therefore reasonable to assume that, if the insects are previously frozen extracellularly at temperatures lower than about $-30°C$, scarcely any water crystallizes within the cells even at super-low temperatures. This assumption seems to be supported by Shinozaki's recent work (1962). He demonstrated in the slug caterpillar, *M. flavescens,* that the amount of water crystallizing at $-30°C$ was more than nine-tenths of the total water content, or nearly all of the freezable water in the insect, and that at lower temperatures, ice increased very gradually (Fig. 17). This was also true in the blood of the larva of a Pyralid moth, *Loxostege sticticalis* (Salt, 1955). Under such conditions the occurrence of intracellular freezing may conceivably be very difficult or impossible, particularly in glycerolated insects.

On the other hand, one of the main causes of injury during cooling in the temperature range below $-30°C$ may probably involve mechanical damage in frozen tissues resulting from the difference of thermal contraction between the tissue and ice crystals, and also between the frozen tissues themselves. In fact the usual stepwise freezing to super-low temperatures, when applied to pupal bodies of the *cecropia* silkworm, resulted in an apparent destruction of the surface cuticle layer and the underlying tissue. In these large insects, of more than 4 to 7 g in weight, a very gradual cooling between $-70°C$ and the liquid nitrogen temperature was observed to be necessary to obtain survival after thawing (Asahina and Tanno, 1966) (Fig. 16).

Despite a fairly large proportion of survival after thawing, the present pre-freezing method has, as a rule, failed to produce a completion of metamorphosis of the treated insects until Tanno's recent success in the prepupa of the sawfly, *T. populi* (Tanno, 1968b). In this insect, likewise, the usual pre-freezing method has been known to result in a failure to shed their pupal cuticles (Tanno and Asahina, 1964). Tanno applied the following special three-step pre-freezing procedure: the prepupae, previously frozen at $-20°C$, were trans-

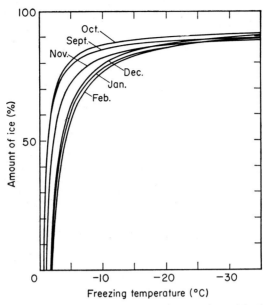

Fig. 17. Seasonal changes in the relative amount of ice formed in the frozen prepupa of *M. flavescens*. Ordinate: percentage of freezable water in the total water content of the insect (from Shinozaki, 1962).

ferred to −5°C and kept there for several hours. These were then slowly re-cooled down to −30°C and were finally transferred directly in liquid nitrogen. After immersion in liquid nitrogen for 24 hours, the insects were rewarmed in the air at room temperature. Of the 20 prepupae examined 15 were able to complete meta-morphosis and appeared on the wing (Tanno, 1968b). The pattern of ice formation in the pre-frozen prepupae was also examined by means of freeze-sectioning (Tanno, 1968b). All the fat body cells, which are the largest cells in the insect, were observed to freeze extracellu-larly. In the prepupae treated with usual one-step pre-freezing, the spaces between the fat body cells were tightly packed with a large number of small ice particles, while in the insects treated with the special three-step pre-freezing a small number of large ice particles with a smooth outline were found between fat body cells. A thicker layer of condensed blood surrounding the large ice particles in the latter insects, than that surrounding the small ice particles in the former insects, was also observed. It was assumed from these results that after the usual one-step pre-freezing some of the fat body cells might be injured by the mechanical stress arising at the thermal contraction of the insect body during cooling to very low temperatures, since the fat body cells were observed to play a very important role in completing metamorphosis (see Section IV, B). The special three-step pre-freezing procedure described above may provide a favourable arrangement of both ice particles and fat body cells to decrease such mechanical stress between them (Tanno, 1968b).

The present pre-freezing method is still insufficient to protect an intact insect at super-low temperatures, for this method cannot always abolish the detrimental effect of severe freezing on further development in the rewarmed insects. However, because of the simplicity of the procedure and its applicability to a wide variety of organisms, the pre-freezing method may be a useful way of preserving life at super-low temperatures, particularly when an adequate protective additive is available.

REFERENCES

Andrewartha, H. G. (1952). Diapause in relation to the ecology of insects. *Biol. Rev.* **27**, 50–107.

Aoki, K. (1956). The undercooling point and frost resistance in the prepupa of a ruby-tailed wasp, *Crysis (Pentacrysis) shanghaiensis. Low Temp. Sci.* Ser. B. **14**, 121–124. (In Japanese.)

Aoki, K. (1962). Protective action of the polyols against freezing injury in the silkworm egg. *Sci. Rep. Tôhoku Univ.,* Ser. IV. **28**, 29–36.

Aoki, K. and Shinozaki, J. (1953). Effect of cooling rate on the undercooling points of the prepupa of slug moth. *Low Temp. Sci.* **10**, 109–115.*

Asahina, E. (1953a). Freezing process of egg cell of sea urchin. *Low Temp. Sci.* **10**, 81–92.*

Asahina, E. (1953b). Freezing process of blood of a frost-hardy caterpillar, *Cnidocampa flavescens. Low Temp. Sci.* **10**, 117–126.*

Asahina, E. (1955). Freezing and supercooling as a method of storage of mobile animal, a preliminary experiment. *Zool. Mag. (Tokyo)* **64**, 280–285.*

Asahina, E. (1956). The freezing process of plant cell. *Contr. Inst. low Temp. Sci. Hokkaido Univ.* **10**, 83–126.

Asahina, E. (1958). On a probable freezing process of molluscan cells enabling them to survive at a super-low temperature. *Low Temp. Sci.* Ser. B. **16**, 65–75.

Asahina, E. (1959a). Prefreezing as a method enabling animals to survive freezing at an extremely low temperature. *Nature, Lond.* **184**, 1003–1004.

Asahina, E. (1959b). Cold-hardiness in overwintering insects. In *Recent Advances in Experimental Morphology.* 92–113 (M. Fukaya, M. Harizuka and K. Takewaki, eds). Yôkendo: Tokyo. (In Japanese.)

Asahina, E. (1959c). Diapause and frost resistance in a slug caterpillar, *Kontyû* **27**, 47–55.

* In Japanese with English summary.

Asahina, E. (1959d). Frost resistance in a nematode *Aphelenchoides ritzema-bosi. Low Temp. Sci.* Ser. B. **17**, 51–62.*

Asahina, E. (1961). Intracellular freezing and frost resistance in egg-cells of the sea urchin. *Nature, Lond.* **191**, 1263–1265.

Asahina, E. (1962). A mechanism to prevent the seeding of intracellular ice from outside in freezing living cells. *Low Temp. Sci.* Ser. B. **20**, 45–56.*

Asahina, E. (1966). Freezing and frost resistance in insects. In *Cryobiology,* 451–486 (H. T. Meryman, ed.). Academic Press, London.

Asahina, E. and Aoki, K. (1958a). Survival of intact insects immersed in liquid oxygen without any antifreeze agent. *Nature, Lond.* **182**, 327–328.

Asahina, E. and Aoki, K. (1958b). A method by which frost-hardy caterpillars survive freezing at a super-low temperature. *Low Temp. Sci.* Ser. B. **16**, 55–63.*

Asahina, E., Aoki, K. and Shinozaki, J. (1954). The freezing process of frost-hardy caterpillars. *Bull. ent. Res.* **45**, 329–339.

Asahina, E. and Takehara, I. (1964). Supplementary notes on the frost resistance in a slug caterpillar, *Monema flavescens. Low Temp. Sci.* Ser. B. **22**, 79–90.*

Asahina, E. and Tanno, K. (1963a). A remarkably rapid increase of frost resistance in fertilized egg cells of the sea urchin. *Expl Cell Res.* **31**, 223–225.

Asahina, E. and Tanno, K. (1963b). A protoplasmic factor of frost resistance in sea urchin egg cells. *Low Temp. Sci.* Ser. B. **21**, 61–69.*

Asahina, E. and Tanno, K. (1964). A large amount of trehalose in a frost-resistant insect. *Nature, Lond.* **204**, 1222.

Asahina, E. and Tanno, K. (1966). Freezing resistance in the diapausing pupa of the *ceropia* silkworm at liquid nitrogen temperature. *Low Temp. Sci.* Ser. B. **24**, 25–34.*

Asahina, E. and Tanno, K. (1967). Observations of ice crystals formed in frozen larvae of the European corn-borer by the use of freeze-sectioning. *Low Temp. Sci.* Ser. B. **25**, 105–111.*

Asahina, E. and Tanno, K. (1968). A frost resistant adult insect, *Pterocormus molitorius* (Hymenoptera, Ichneumonidae). *Low Temp. Sci.* Ser. B. **26**, 85–89.*

Bachmetjew, P. (1901). *Experimentelle entomologische Studien vom physikalisch-chemischen Standpunkt aus,* **1**, 160 pp. Leipzig.

Barnes, D. and Hodson, A. C. (1956). Low temperature tolerance of the European corn borer in relation to winter survival in Minnesota. *J. econ. Ent.* **49**, 19–24.

Chambers, R. and Hale, H. P. (1932). The formation of ice in protoplasm. *Proc. R. Soc.* Ser. B. **110**, 336–352.

Chino, H. (1957). Conversion of glycogen to sorbitol and glycerol in the diapause egg of the bombyx silkworm. *Nature, Lond.* **180**, 606–607.

Chino, H. (1958). Carbohydrate metabolism in the diapause egg of the silkworm, *Bombyx mori.* II. Conversion of glycogen into sorbitol and glycerol during diapause. *J. Insect Physiol.* **2**, 1–12.

De Coninck, L. (1951). On the resistance of the free-living nematode *Anguillula silusiae* to low temperatures. *Biodynamica* **7**, 77–84.

* In Japanese with English summary.

Doebbler, G. F., Rowe, A. W. and Rinfret, A. P. (1966). Freezing of mammalian blood and its constituents. In *Cryobiology*, 407–450 (H. T. Meryman, ed.). Academic Press, London.

Ditman, L. P., Vogt, G. B. and Smith, D. R. (1942). The relation of unfreezable water to cold-hardiness of insects. *J. econ. Ent.* **35**, 265–272.

Dorsey, N. E. (1948). The freezing of supercooled water. *Trans. Am. phil. Soc.* **38**, 247–328.

Dubach, P., Smith, F., Pratt, D. and Stewart, C. M. (1959). Possible role of glycerol in the winter-hardiness of insects. *Nature, Lond.* **184**, 288–289.

Duchâteau, G., Florkin, M. and Leclercq, J. (1953). Concentrations des bases fixes et types de composition de la base totale de l'hémolymphe des insectes. *Archs int. Physiol.* **61**, 518–549.

Fukaya, M. and Mitsuhashi, J. (1961). Larval diapause in the rice stem borer with special reference to its hormonal mechanisms. *Bull. natn. Inst. agric. Sci., Tokyo* C. 13, 1–32.

Hanec, W. and Beck, S. D. (1960). Cold hardiness in the European corn borer, *Pyrausta nubilalis* (Hübn.). *J. Insect. Physiol.* **5**, 169–180.

Hinton, H. E. (1960). Cryptobiosis in the larva of *Polypedilum vanderplanki* Hinton (Chironomidae). *J. Insect Physiol.* **5**, 286–300.

Kistler, S. S. (1936). The measurement of "bound" water by the freezing method. *J. Am. chem. Soc.* **58**, 901–907.

Kozhantshikov, I. W. (1938). Physiological conditions of cold-hardiness in insects. *Bull ent. Res.* **29**, 253–262.

Leader, J. P. (1962). Tolerance to freezing of hydrated and partially hydrated larvae of *Polypedilum* (Chironomidae). *J. Insect Physiol.* **8**, 155–163.

Levitt, J. (1962). A sulfhydryl-disulfide hypothesis of frost injury and resistance in plants. *J. Theoret. Biol.* **3**, 355–391.

Losina-Losinsky, L. K. (1937). Cold-hardiness and anabiosis in larvae of *Phyrausta nubilalis*. *Zool. Zh.* **16**, 614–642. (In Russian with English summary.)

Losina-Losinsky, L. K. (1962). Survival of insect at super-low temperatures. *Dokl, Akad. Nauk SSSR* **147**, 1247–1249. (In Russian.)

Losina-Losinsky, L. K. (1963). Resistance of some insects to the temperature of liquid helium ($-269°$C) under conditions of intracellular freezing in absence of antifreezes. *Cytologya, Akad. Nauk* **5**, 220–221. (In Russian.)

Losina-Losinsky, L. K. (1965). Survival of some insects and cells following intracellular ice formation. *Fedn Proc.* **24**, Suppl. 15, 206–211.

Lovelock, J. E. (1953a). The haemolysis of human red blood cells by freezing and thawing. *Biochim. biophys. Acta* **10**, 414–426.

Lovelock, J. E. (1953b). The mechanism of protective action of glycerol against haemolysis by freezing and thawing. *Biochim. biophys. Acta* **11**, 28–36.

Luyet, B. J. and Gehenio, P. M. (1940). *Life and Death at Low Temperatures*, 341 pp. Biodynamica: Normandy, Missouri.

Luyet, B. J. and Gibbs, M. C. (1937). On the mechanism of congelation and of death in the rapid freezing of epidermal plant cells. *Biodynamica* **25**, 1–18.

Luyet, B. J. and Keane, J. Jr. (1955). A critical temperature range apparently characterized by sensitivity of bull semen to high freezing velocity. *Biodynamica* **7**, 281–292.

* In Japanese with English summary.

Mazur, P. (1963). Kinetics of water loss from cells at subzero temperatures and the likelihood of intracellular freezing. *J. gen. Physiol.* **47**, 347–369.

Meryman, H. T. (1956). Mechanics of freezing in living cells and tissues. *Science, N.Y.* **124**, 515–521.

Ôura, H. (1950). On the rate of nucleation of ice. *J. phys. Soc. Japan* **5**, 277–279.

Payne, N. M. (1927). Freezing and survival of insects at low temperatures. *J. Morph.* **43**, 521–546.

Robinson, W. (1928). Relation of hydrophilic colloids to winter hardiness of insects. *Colloid Symp. Monog.* **5**, 199–218.

Sacharov, N. L. (1930). Studies in cold resistance of insects. *Ecology* **11**, 505–517.

Sakai, A. (1956). Survival of plant tissue at super-low temperatures I. *Low Temp. Sci.* Ser. B. **14**, 17–23.*

Sakai, A. (1962). Studies on the frost-hardiness of woody plants I. The causal relation between sugar content and frost-hardiness. *Contr. Inst. low Temp. Sci. Hokkaido Univ.* Ser. B. **11**, 1–40.

Salt, R. W. (1936). Studies on the freezing process in insects. *Tech. Bull. Minn. agric. Exp. Stn* **116**, 41 pp.

Salt, R. W. (1953). The influence of food on the cold-hardiness of insects. *Can. Ent.* **85**, 261–269.

Salt, R. W. (1955). Extent of ice formation in frozen tissues, and a new method for its measurement. *Can. J. Zool.* **33**, 391–403.

Salt, R. W. (1956). Influence of moisture content and temperature on cold-hardiness of hibernating insects. *Can. J. Zool.* **34**, 283–294.

Salt, R. W. (1957). Natural occurrence of glycerol in insects and its relation to their ability to survive freezing. *Can. Ent.* **89**, 491–494.

Salt, R. W. (1958a). Application of nucleation theory to the freezing of supercooled insects. *J. Insect Physiol.* **2**, 178–188.

Salt, R. W. (1958b). Relationship of respiration rate to temperature in a supercooled insect. *Can. J. Zool.* **36**, 265–268.

Salt, R. W. (1958c). Role of glycerol in producing abnormally low supercooling and freezing points in an insect, *Bracon cephi* (Gehan). *Nature, Lond.* **181**, 1281.

Salt, R. W. (1959a). Survival of frozen fat body cells in an insect. *Nature, Lond.* **184**, 1426.

Salt, R. W. (1959b). Role of glycerol in the cold-hardiness of *Bracon cephi* (Gehan). *Can. J. Zool.* **37**, 59–69.

Salt, R. W. (1961a). Principles of insect cold-hardiness. *A. Rev. Ent.* **6**, 55–74.

Salt, R. W. (1961b). A comparison of injury and survival of larvae of *Cephus cinctus* Nort. after intracellular and extracellular freezing. *Can. J. Zool.* **39**, 349–357.

Salt, R. W. (1962). Intracellular freezing in insects. *Nature, Lond.* **193**, 1207–1208.

Salt, R. W. (1963). Delayed inoculative freezing of insects. *Can. Ent.* **95**, 1190–1202.

Salt, R. W. (1966a). Factors influencing nucleation in supercooled insects. *Can. J. Zool.* **44**, 117–133.

* In Japanese with English summary.

Salt, R. W. (1966b). Effect of cooling rate on the freezing temperatures of supercooled insects. *Can. J. Zool.* **44**, 655–659.

Salt, R. W. (1966c). Relation between time of freezing and temperature in supercooled larvae of *Cephus cinctus* Nort. *Can. J. Zool.* **44**, 947–952.

Scholander, P. F., Flagg, W., Hock, R. J. and Irving, L. (1953). Studies on the physiology of frozen plants and animals in the Arctic. *J. cell comp. Physiol.* **42**, Suppl. 1, 1–56.

Shinozaki, J. (1954a). The velocity of crystallization of ice from the blood of prepupae of slug moth. *Low Temp. Sci.* Ser. B. **11**, 1–11.*

Shinozaki, J. (1954b). On the freezing of the prepupae of slug moth. *Low Temp. Sci.* Ser. B. **12**, 71–86.*

Shinozaki, J. (1962). Amount of ice formed in the prepupa of slug moth and its periodicity. *Contr. Inst. low Temp. Sci. Hokkaido Univ.* Ser. B. **12**, 1–52.

Smith, A. U. (1954). Effect of low temperatures on living cells and tissues. In *Biological Applications of Freezing and Drying*, 1–50 (R. I. C. Harris, ed.). Academic Press, New York.

Smith, A. U. (1961). *Biological Effects of Freezing and Supercooling*, 462 pp. Edward Arnold Ltd., London.

Smith-Johansen, R. (1948). Some experiments in the freezing of water. *Science, N.Y.* **108**, 652–654.

Sømme, L. (1964). Effects of glycerol on cold-hardiness in insects. *Can. J. Zool.* **42**, 87–101.

Sømme, L. (1965). Further observations on glycerol and cold-hardiness in insects. *Can. J. Zool.* **43**, 765–884.

Stobbart, R. H. and Shaw, J. (1964). Salt and water balance: Excretion. In *The Physiology of Insecta* III, 189–258 (M. Rockstein, ed.). Academic Press, New York.

Takehara, I. (1963). Glycerol in a slug caterpillar II. Effect of some reagents on glycerol formation. *Low Temp. Sci.* Ser. B. **21**, 55–60.*

Takehara, I. (1966). Natural occurrence of glycerol in the slug caterpillar, *Monema flavescens. Contr. Inst. low Temp. Sci. Hokkaido Univ.* Ser. B. **14**, 1–34.

Takehara, I. and Asahina, E. (1959). Glycerol Content in some frost-hardy insects, a preliminary report. *Low Temp. Sci.* Ser. B. **17**, 159–163.*

Takehara, I. and Asahina, E. (1960a). Glycerol in the overwintering prepupa of slug moth, a preliminary note. *Low Temp. Sci.* Ser. B. **18**, 51–56.*

Takehara, I. and Asahina, E. (1960b). Frost resistance and glycerol content in overwintering insects. *Low Temp. Sci.* Ser. B. **18**, 57–65.*

Takehara, I. and Asahina, E. (1961). Glycerol in a slug caterpillar I. Glycerol formation, diapause and frost-resistance in insect reared at various graded temperatures. *Low Temp. Sci.* Ser. B. **19**, 29–36.*

Tanno, K. (1962). Frost resistance in a carpenter ant *Camponotus obscuripes obscuripes*. I. The relation of glycerol to frost resistance. *Low Temp. Sci.* Ser. B. **20**, 25–34.*

Tanno, K. (1963). Frost resistance in overwintering pupa of a butterfly *Papilio xuthus. Low Temp. Sci.* Ser. B. **21**. 41–43.*

Tanno, K. (1964). High sugar levels in the solitary bee, *Ceratina. Low Temp. Sci.* Ser. B. **22**, 51–57.*

* In Japanese with English summary.

Tanno, K. (1965a). The fat-cell in the sawfly, *Trichiocampus populi* Okamoto. *Low Temp. Sci.* Ser. B. **23**, 37–45.*

Tanno, K. (1965b). Frost resistance in the poplar sawfly, *Trichiocampus populi* Okamoto. II. Extracellular and intracellular freezing in fat-cells. *Low Temp. Sci.* Ser. B. **23**, 47–53.*

Tanno, K. (1965c). Frost resistance in the poplar sawfly, *Trichiocampus populi* Okamoto. III. Frost resistance and sugar content. *Low Temp. Sci.* B. **23**, 55–64.*

Tanno, K. (1967a). Freezing injury in fat-body cells of the poplar sawfly. In *Cellular Injury and Resistance in Freezing Organisms*, 245–257 (E. Asahina, ed.). Institute of Low Temperature Science: Sapporo.

Tanno, K. (1967b). Immediate termination of prepupal diapause in poplar sawflies by body freezing. *Low Temp. Sci.* Ser. B. **25**, 97–103.*

Tanno, K. (1968a). Frost resistance in the poplar sawfly, *Trichiocampus populi* Okamoto. IV. Intracellular freezing in fat-cells and injury occurring upon metamorphosis. *Low Temp. Sci.* Ser. B. **26**, 71–78.*

Tanno, K. (1968b). Frost resistance in the poplar sawfly, *Trichiocampus populi* Okamoto. V. Freezing injury at the liquid nitrogen temperature. *Low Temp. Sci.* Ser. B. **26**, 79–84.*

Tanno, K. and Asahina, E. (1964). Frost resistance in the poplar sawfly, *Trichiocampus populi* Okamoto. *Low Temp. Sci.* Ser. B. **22**, 59–70.*

Ushatinskaya, R. S. (1957). *Principles of Cold Resistance in Insects*, 314 pp. Academy of Sciences U.S.S.R. Press: Moscow. (In Russian.)

Uvarov, B. P. (1931). Insect and climate. *Trans. R. ent. Soc. Lond.* **79**, 1–247.

Way, M. J. (1960). The effects of freezing temperatures on the developing egg of *Leptohylemyia coarctata* Fall. (Diptera, Musscidae) with special reference to diapause development. *J. Insect Physiol.* **4**, 92–101.

Wigglesworth, V. B. (1964). The hormonal regulation of growth and reproduction in insects. In *Advances in Insect Physiology* **2**, 247–336 (J. W. L. Beament, J. E. Treherne and V. B. Wigglesworth, eds.). Academic Press, London.

Wigglesworth, V. B. (1965). *The Principles of Insect Physiology.* 6th edn. 741 pp. Methuen: London.

Wilbur, K. M. and McMahan, E. A. (1958). Low temperature studies on the isolated heart of the beetle, *Popilius disjunctus* (Illiger). *Ann ent. Soc. Am.* **51**, 27–32.

Wyatt, G. R. (1967). The biochemistry of sugars and polysaccharides in Insects. In *Advances in Insect Physiology* **4**, 287–360 (J. W. L. Beament, J. E. Treherne and V. B. Wigglesworth, eds). Academic Press, London.

Wyatt, G. R. and Meyer, W. L. (1959). The chemistry of insect hemolymph. III. Glycerol. *J. gen. Physiol.* **42**, 1005–1011.

* In Japanese with English summary.

Neural Control of Firefly Luminescence*

A. D. CARLSON

Department of Biological Sciences
State University of New York at Stony Brook
Stony Brook, New York, U.S.A.

I. INTRODUCTION

The flash of the adult firefly, a brilliant burst of light produced by the coordinated participation of hundreds of units, is an outstanding example of an insectan effector response. Because it is so easily observed it provides an excellent tool for the analysis of neural function in an insect. Study of the mechanism of flash control over the last 20 years has revealed a neuroeffector system which contains a number of unique features.

*The original work reported here was supported by NSF grant GB-6385. Thanks are due to C. Kaars for permission to produce previously unpublished figures.

The spontaneous flash of the adult firefly has been conclusively shown to be the mechanism by which the sexes attain proximity (Osten-Sacken, 1861; McDermott, 1911; Mast, 1912). Among the species making up the genera *Photuris* and *Photinus* there is considerable variation in flash duration, which ranges from about 150 msec to over 1000 msec, and in flash pattern, which can be a smooth rounded pulse or a rapid series of twinkles. The adult firefly is also capable of producing a wide range of qualitatively different light responses. These include glows of varying intensity and duration, and scintillation, which is the rapid, uncoordinated flashing of hundreds of individual elements. Artificially a pseudoflash can be induced by the inrush of oxygen into an hypoxic lantern (Fig. 1A).

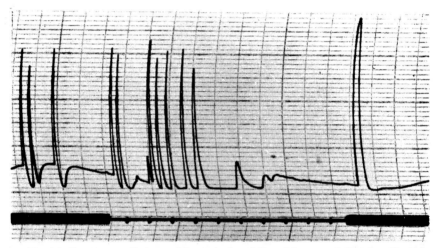

Fig. 1A. Spontaneous flashes and hypoxic glow followed by a pseudoflash in a *Photuris,* male. Upper trace is photomultiplier output; middle trace heavy line is 21% oxygen and narrow line is nitrogen; marks on narrow segment are time base, 1 mark per second, reading from left to right. Hypoxic glow begins approximately 7 seconds after hypoxic onset; note its relatively low intensity. (From Carlson, 1965.)

The larval firefly, containing tiny lanterns of a structure different from the adult, and of uncertain function, typically produces a glow which lasts for seconds. Its lanterns are incapable of flashing (Fig. 1B).

The control of firefly luminescence has been the subject of intensive study for considerably over one hundred years and a critical account of the earlier work on the physiology of control can be found in Buck (1948). Up to that time two principal theories of

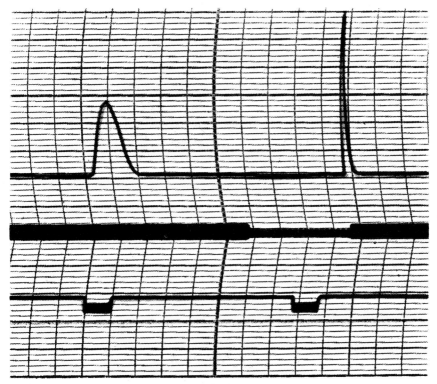

Fig. 1B. Stimulated glow and pseudoflash of *Photuris* larva. Upper and middle traces same as upper and lower traces respectively in Fig. 1A. Bottom trace: stimulus, 5 volts, 20 msec duration, 10 per second frequency. Electrode pair inserted in 6th abdominal segment. (From Carlson, 1965.)

neural control had emerged. One theory advanced by Snell (1932) and supported by Alexander (1943) held that the nerves regulated the opening of a valve in the tracheal end cell which controlled the inflow of oxygen, a chemical known to be required in light production. This theory was critically examined by Buck (1948) who seriously questioned the physiological basis for such a mechanism of oxygen control of luminescence. Not only did it raise the problem of maintaining the light producing cells virtually anaerobic during the long non-flashing periods, but the known mechanics of oxygen diffusion in insects could not support the kinetics of the flash response. The discovery by Beams and Anderson (1955) that the "fibers" supposedly involved in the valve were, in reality, mitochondria removed even the morphological basis of the theory. Buck

(1948, 1955) supported a theory that all the reactants necessary for luminescence were present within the photocyte, the light producing cell, and were released to react by neural activity. Buck suggested that neural influence might act through control at the enzyme level. McElroy (1951) strengthened this proposal by demonstrating that a reasonable flash could be produced *in vitro* by freeing the enzyme, luciferase, from an inactive complex. The current general theory is that neural activity. causes the indirect release of luciferase which reacts with the necessary substrates, all of which are present in excess, the flash terminating when the enzyme is again sequestered in an inactive complex.

This review will focus on the physiological basis of three related aspects of the general problem of neural control. It will describe in some detail the chain of neural events which begin in the brain and culminate in the light response. This represents the basic machinery of neural control of luminescence. Once begun, this process normally culminates in a stereotyped effect, but central and peripheral neural modulating mechanisms do exist which control flash frequency and intensity. This represents a second aspect of neural control. Finally, behavioral analysis of brain function involved in signal and species discrimination and in synchronous flashing will be treated as a third aspect of neural control.

Only those aspects of lantern morphology, luminescence biochemistry and development which contribute directly to an understanding of neural control mechanisms will be included. Various aspects of the firefly luminescence have already been treated in admirable fashion and the reader is referred to the following: historical background on the physiology of control (Buck, 1948), biochemistry of adult luminescence (McElroy, 1964), morphology (Smith, 1963), Taxonomy, behavior and ecology of *Photinus* (Lloyd, 1966).

II. ANATOMY OF THE FIREFLY LANTERN

A. ADULT LANTERN

By 1948 most of the major structural features of the lanterns of the genera *Photuris* and *Photinus* could be summarized by Buck from the work of a number of investigators (McDermott and Crane, 1911; Dahlgren, 1917; Seaman, 1891; Lund, 1911; Williams, 1916; Hess, 1922; Townsend, 1904).

The lantern of the adult male occupies the ventral portion of the 6th and 7th abdominal segments. It is composed of a ventral photogenic layer overlaid by a dorsal reflector layer. Trachea run vertically through both layers surrounded by tracheal epithelial cells which, in the photogenic layer, form cylindrical enlargements. In this layer the trachea give off tracheoles. Projecting away from the cylinder, the fine tracheoles are surrounded by a tracheal end cell which closely resembles the tracheal epithelial cell against which it abuts. The tracheal end cells are filled with relatively long mitochondria (Beams and Anderson, 1955). In the end cell region the tracheoles split into two or more branches still ensheathed by tracheolar cells. These cells contain mitochondria arranged along the membrane adjacent to the tracheole. The photocytes surround the tracheal trunks and their associated tracheal end cells in rosette fashion. They contain granules, called photocyte granules which have characteristic vesicular compartments connected with the exterior by a long tube (Smith, 1963). A cortical zone, free of photocyte granules, exists where the photocytes abut the tracheolar cells. This differentiated zone does contain large numbers of short, rod-shaped mitochondria (Beams and Anderson, 1955).

The ventral nerve cord supplies the still more ventral lying lantern principally from the last two abdominal ganglia. These are both situated in segment VI. Hanson (1962) examined the gross innervation of the lantern by means of stimulation and nerve transection studies. The scheme of innervation he developed is shown in Fig. 2. In gross innervation luminous segments do not differ from non-luminous segments except for having larger segmental nerves due to the added function of the light organ. Hanson was able to demonstrate that lantern luminescence is controlled by peripheral nerves which supply morphologically circumscribed areas. Weak stimulation of a small region of photogenic tissue or the peripheral nerve innervating that same region resulted in activation of that area only. Under these conditions the organ did not glow at the point of electrode insertion unless the nerve controlling that point was excited. Stimulation at higher intensities activated the entire side of the segment due to recruitment. Intersegmental linkages on the lateral sides of both photogenic segments existed and the possibility arose that these areas could be innervated from two ganglia. It could not be determined, however, whether each ganglion controlled minute but discrete areas or dual innervation of individual effector units existed. When one trunk of the nerve cord between the 5th and 6th abdominal

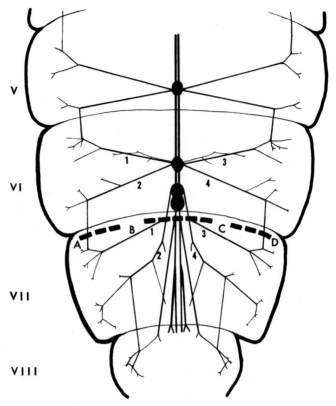

Fig. 2. Dorsal view of posterior abdomen. Roman numerals refer to segments, numbers designate nerves. Broken line ABCD locates the incisions used in denervation experiments. (From Hanson, 1962.)

ganglia was severed and only the intact trunk was stimulated, the entire lantern responded. This result demonstrated that decussation of motor pathways existed in the lantern ganglia.

Up to this point, the study of neural control suffered the embarrassment of not knowing where the lantern nerves terminated. Kluss (1958) showed that nerve trunks accompanied the trachea into the cylinder. We owe to Smith (1963) the resolution of the detailed neural anatomy of the *Photuris* lantern, which is illustrated in Fig. 3. He showed that the lantern nerves split out of the nerve sheath near the level of the tracheole, crease the tracheal end cell and terminate in pad-like endings between the end cell and tracheolar cell (Fig. 4). The gaps between the plasma membranes of the nerve

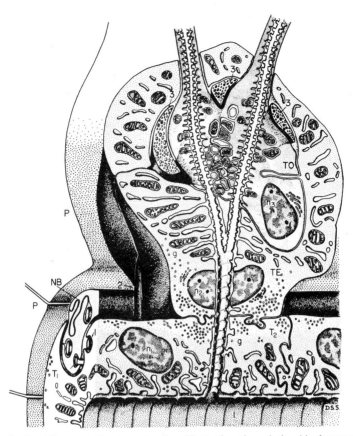

Fig. 3. Semidiagrammatic reconstruction illustrating the relationship between nerve tissue and the cells of the tracheal system in the lantern of *P. pennsylvanica* based on electron micrographs. Portions of two tracheal epithelial cells (T_1, T_2) are shown, sectioned transversely at lower left and longitudinally elsewhere: the margins of transversely sectioned photocytes (P) are indicated along the left of the figure. The cells of the tracheal trunk surround the cuticle-limited lumen (l), a lateral branch of which divides into two. Shortly afterwards the tracheolar tubes thus formed leave the tracheal end cell (TE) and enter the tracheolar cell (TO), and cytoplasmic processes of the latter surround the tracheolar branches as they emerge from the encompassing end cell to penetrate between the photocytes. The complex folding of the tracheolar cell membrane around the tracheal lumen and deposits believed to be glycogen (g) in the tracheal epithelial cells and the end cell are shown. The nuclei of the tracheal epithelial cell, the end cell and the tracheolar cell are labeled n_1, n_2 and n_3 respectively. A peripheral nerve branch (NB) is shown in transverse section indenting the tracheal epithelial cell at lower left. An axon becomes freed from the lemnoblast (arrowed 1) and follows a lateral course alongside the end cell (arrowed 2) and ends in dilated, vesicle-packed "terminal processes" (arrowed 3) around the "cell body" of the tracheolar cell, and tightly inserted between the surface of the latter and the concave distal portion of the end cell. This figure is intended to indicate only the spatial relations between the cells, rather than the relative size of the cells and of their cytological components. In particular the size, relative to that of the end cell, of the terminal processes of the axon and of the tracheolar cell processes has been exaggerated for clarity, and the actual proportions may be seen in Fig. 4. (From Smith, 1963.)

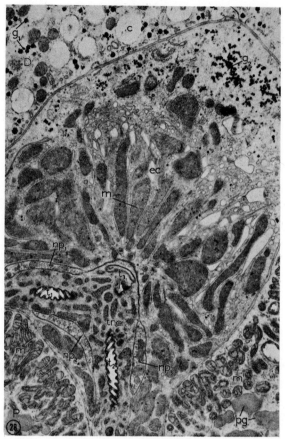

Fig. 4. A survey field representing an area of junction between cells of the dorsal layer
(D) and the photocyte epithelium (P), including profiles of a tracheal end cell (ec) and a
tracheolar cell (trc). Note the concentration of photocyte mitochondria (m') along the
margin of the end cell, and also the photocyte granules (pg). Empty cavities in the dorsal
layer cells (c) are thought to represent a soluble material lost during preparation of the
material: the deeply "stained" deposits in these cells (g_1) are believed to be of glycogen,
and smaller clusters of similar appearance occur in the cytoplasm of the end cell (g_2). Much
of the cytoplasm of the end cell (occupying the bulk of this field) is filled with more or less
radially arranged mitochondria (m) situated within a complex system of membrane-limited
processes. The plane of this second passes through the end cell distally both to the point of
branching (in this instance trifurcation) of the lateral tracheal twig (cf. Fig. 3) and also to the
cell body of the tracheolar cell, though processes of the latter, including the tracheolar tubes,
are seen. Three tracheolar branches are present: two of these (t_1, t_2) are sectioned obliquely,
and the third (t_3), transversely. The cytoplasm of the tracheolar cell (trc) contains mito-
chrondria, lying in pockets defined by the membrane surrounding the tracheolar tube, and at
lower center a tracheolar cell process is seen as it is about to leave the confines of the end
cell, to pass between the photocytes. Three terminal nerve processes (np_1, np_2, np_3) are seen
in transverse profile, insinuated between the end cell and tracheolar cell surfaces. $\times$ 16,000.
(From Smith, 1963.)

endings and the membranes of the opposed tracheal end cell and tracheolar cell are about 100Å. It was not possible to determine whether the nerve activity acts across the nerve-tracheal end cell junction, the nerve-tracheolar cell junction, or both.

While in *Photuris* the lantern nerve endings apparently are separated from the photocytes by the thickness of the tracheolar cell. This is not the only arrangement possible: nerve endings have been found to contact the photocytes as well as the tracheolar and tracheal end cells in the Asiatic fireflies, *Pteroptyx, Luciola* and *Pyrophanes* (Peterson and Buck, 1968). Still, at least in *Photuris,* neural-luminescence coupling must occur through either the intervening end cell or the tracheolar cell (Smith, 1963).

Two different populations of vesicle-like inclusions are found in the nerve terminals, beginning near the point where the nerve loses its sheath. One group is 200–400Å in diameter with clear centers; the other group is 600–1200Å in diameter with electron opaque centers. Smith suggests that the large vesicles may have a neurosecretory function. He points out that the complex of two types of vesicles resembles that found by Palay (1957) in terminations within the vertebrate neurohypophysis. The firefly is apparently not unusual in possessing more than one type of vesicle in its neural tissue. This has been shown to occur in fibers of neuropile in the blowfly (Chiarodo, 1968) and in the roach (Hess, 1958; Smith, 1965).

B. LARVAL LANTERN

The tiny, paired, larval lanterns lie on the ventro-lateral surfaces of the 8th abdominal segment. They are anatomically much simpler than the adult lantern. They are composed of a dorsal "reflector" and ventral photogenic layer with a diffuse and unoriented tracheal system. As in the adult, the tracheoles are surrounded by tracheolar cells, but tracheal end cells are absent. The nerves apparently terminate directly on the photocytes in similar vesicle filled endings (M. Wetzel, personal communication). The photocytes are similar to those found in the adult.

III. BIOCHEMISTRY OF THE LIGHT REACTION

The biochemical basis of firefly luminescence has been studied intensively by McElroy and his associates. The general scheme of *in vitro* biochemical interaction and a proposed mechanism of neural control is shown in Fig. 5.

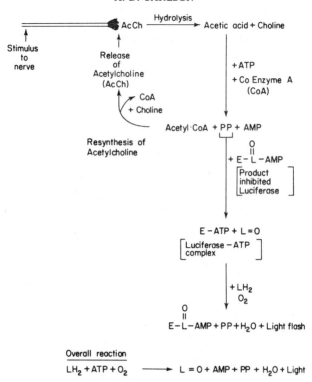

Fig. 5. Suggested biochemical scheme for the control of the firefly flash. See text for description of the reactions. (From McElroy, 1964.)

The enzyme, luciferase (E), in complex with ATP (E-ATP) combines with reduced luciferin (LH_2) to form a complex (E-LH_2-AMP). This is oxidized by molecular oxygen to an oxidized complex (E-L = O-AMP) with the emission of light. The oxidized complex is very stable and, because of this, the luminescent reaction is strongly product inhibited (McElroy, 1957; McElroy and Strehler, 1949; McElroy and Coulombre, 1952).

Luminescence may be controlled by pyrophosphate which can free luciferase from the stable, oxidized complex, thereby allowing it to interact with more reduced luciferin and ATP (McElroy, 1951). Pyrophosphate, produced during formation of the active complex, is destroyed by inorganic pyrophosphatase which exists in the lantern at concentrations in excess of 10 times those in the rest of the body (McElroy *et al.*, 1951; McElroy, *et al.*, 1953). McElroy has suggested that the nerves might trigger luminescence by release of acetylcholine

(AcCh) which interacts with coenzyme A (CoA) to form Acetyl.CoA. This interacts, in turn, with ATP to release pyrophosphate (PP) within the photocytes which frees luciferase from the oxidized complex (McElroy and Hastings, 1955, 1957).

There are a number of objections to the theory of control advanced by McElroy, which he has not hesitated to admit. Acetylcholine has not been found in insect neuromuscular junctions but has been demonstrated in insect C.N.S. It does not induce glowing in lanterns whether injected into the animal or perfused over the isolated lantern. McElroy and Seliger (1965) also point out that the observations of Smith (1963) present a serious objection to the idea of transmitter passing from the axon directly to the photocytes. Smith has suggested two possibilities: (1) Nerve stimulation induces lumine-scence by release of a substance which reaches the photocyte via the end cell, cytoplasm. (2) Transmitter released by nerve impulses may depolarize the tracheolar cell membrane, which in turn triggers the photocytes. In spite of these uncertainties, pyrophosphate has been strongly implicated in the triggering process.

IV. NEURAL INVOLVEMENT IN THE LIGHT REACTION

A. THE ADULT LANTERN AS A NEUROEFFECTOR

The studies of Case and Buck have established the adult lantern as a true neuroeffector organ with many similarities to neuromuscular systems (Buck and Case, 1961). It shows such analogous relation-ships as strength-duration curves, facilitation, summation, constant stimulus-response latency, treppe, tetanus, adaptation and fatigue.

The adult lantern produces a strength-duration relationship similar to insect muscle. *Photuris* shows a 3.9 msec chronaxie and a 2.1 volt rheobase. Response threshold increases of the order of 10-fold between temperatures of 25°C and 10°C.

Chang (1956) initially demonstrated the close correspondence between the adult lantern and neuromuscular systems in terms of summation and treppe. This has been amply confirmed by Buck and Case (1961). They showed that repetitive stimulation at subliminal intensity induces summation with the initial succeeding flashes increasing in intensity (Fig. 6). The facilitation produced by paired pulses of varying interval is shown in Figs 7 and 8. The minimum facilitation delay to produce perceptible summation with near threshold stimuli is 5 msec in *Photuris* and 6–8 msec in *Photinus pyralis.*

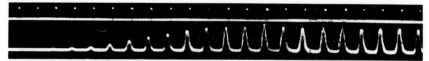

Fig. 6. Summation of flash responses recorded by photomultiplier in a decapitated *Photuris* firefly, sex unrecorded, to stimuli of 4 msec duration, 5 volts and 10 per second. Electrodes were in lantern. Total duration of oscilloscope record is 2.2 seconds. (From Buck and Case, 1961.)

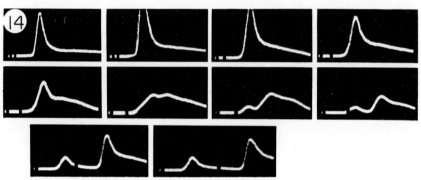

Fig. 7. Facilitation to paired shocks of increasing interval in a decapitated *Photuris* female. Electrodes were in lantern. Paired stimuli were 6 msec duration and 15 volts with intervals of 15, 20, 30, 40, 50, 60, 70, 90, 150 and 200 msec. Duration of oscilloscope frames were 360 msec except last two of 430 msec. Stimulus indicated by break in photomultiplier trace. (From Buck and Case, 1961.)

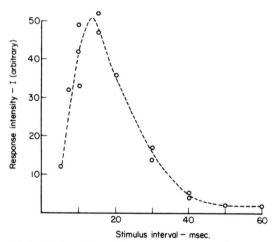

Fig. 8. Temporal facilitation of electrically induced flashes. Abscissa represents interval from first to second stimulus of a pair, in all instances 3 msec duration and 5 volts. Woods Hole *Photuris* female, isolated lantern. Typical of five other specimens. (From Buck and Case, 1961.)

The frequency response of the flash is, to a large extent, dependent on the species flash duration. *Photuris* could respond with completely separate flashes to stimulus frequencies of 10 per second and could give 1 : 1 peak responses as high as 20 per second. *Photinus pyralis* could produce separate flashes only up to 1 per second with 1 : 1 peak response up to 4 per second due to its inherently longer flash. At higher stimulation frequencies, flashes began to merge and the result was found to be similar to tetanus in striated muscle.

Animals were highly variable with respect to adaptation and fatigue. Buck and Case (1961) have observed one *Photuris* female to flash once per second for over 70 minutes while the response faded in other adults after only a few stimuli. Response loss was particularly prevalent after bouts of intense stimulation, which suggested it might be due to junctional fatigue.

Stimulus response latency was found to be remarkably constant within a particular species but it was also extremely long when compared to latencies found in typical neuromuscular systems. Latency to stimuli from electrodes placed directly in the lantern tissue at 22°C averaged 86 msec in *Photuris,* and 194 msec in the Maryland *Photinus pyralis* tested. Less than 5% variation in delay time was found in lanterns stimulated with identical consecutive pulses. Latency was found to be temperature sensitive, showing a Q_{10} of over 2.0 at 20°C and of about 1.5 at 30°C.

By stimulating the lantern directly, Case and Buck observed that as the stimulation intensity, or duration, was increased, a flash of significantly shorter latency appeared, followed by the normal flash at its typical latency. They have named this early response the "quick" flash and the later flash, induced at lower stimulation intensities, the "slow" flash (Fig. 9). More extensive examination of this phenomenon by Buck *et al.* (1963) revealed that the quick flash latency was inversely related to stimulus intensity and duration; it

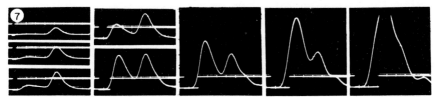

Fig. 9. Development of the quick flash in the isolated abdomen of an Iowa *Photuris,* male. Eight successive responses to stimulus durations of 5, 7.5, 10, 15, 20, 30, 40 and 50 msec respectively; voltage probably 150 volts. Oscilloscope frames equal 175 msec. (From Buck *et al.* 1963).

A. D. CARLSON

reached a lower limit of 18 msec at high stimulation intensities (Table I). While the quick flash appeared to involve all of the lantern

Table I

Approximate peripheral response latencies (msec; 25°)

Species	Ultra-short	Quick flash	Slow flash	Ultra-long
Woods Hole *Photuris*	1	18	70	
Maryland *Photinus pyralis*		20	160	
Photinus punctulatus		15	75, 110*	
Photinus consanguineus		10	100, 175†, 250†	
Photinus marginellus		5–10	40–200	
Md. and Iowa *Photuris* larva	1–3			600

* Differs with sex.
† Secondary peaks, not appearing alone.
(From Buck *et al.* 1963.)

surface, diffusion of light within the lantern made this conclusion uncertain. Quick flashes were less stable, but the quick and slow flashes resembled each other closely in form and kinetics, which suggested that they involved the same basic effector events. It could not be determined however, whether an individual photocyte could produce consecutive quick and slow flashes. The observation that the quick and slow flashes do not change together, and that one is usually enhanced at the expense of the other, suggests that the photocytes are unable to do so.

The *Photuris* quick flash threshold is considerably higher than the slow flash. It has a 10 msec chronaxie and a rheobase at 30 volts compared with 3.9 msec and 2.1 volts respectively for the slow. While the slow flash cannot be induced below 8°C, the quick flash can be induced down to at least 3.5°C. The latency of the quick flash was found to be virtually temperature independent. The quick flash augments with repetitive stimulation and undergoes tetanus (fusion). The intershock interval for best facilitation is 10–15 msec which is identical to that of the slow flash. The limit of 1 : 1 frequency response for the quick flash is over 50 per second in *Photuris* and about 13 per second in *Photinus pyralis*.

Much the same system has been shown to occur in *Luciola lusitanica* by Buonamici and Magni (1967). The flash response was found to conform to the general pattern of neural control shown for *Photuris* and *Photinus*. The lantern, occupying the 6th and 7th abdominal segments, gives typical facilitation and tetanus responses to direct lantern stimulation. Direct stimulation of the lantern induces flashes of 100–160 msec latency, called peripheral flash, but no quick flash could be obtained. More variable responses of 200–750 msec latency, called reflex flashes, do occur, however. This response disappears in decapitated animals pointing to indirect brain stimulation of the lantern via activation of some afferent pathway.

An ultrashort flash of about 1 msec latency was obtained by Buck *et al.* (1963) by stimulating extirpated *Photuris* lanterns or portions of lanterns with closely opposed electrodes (1 mm) at high intensity. It is questionable whether this response represents a truly different excitation pathway. Because of the short latency and its merger with the quick flash, the authors could not observe it visually. Compared with sparks generated on saline moistened filter paper, it was slower, required longer stimulus durations and did not splutter at higher durations. Still, the authors admit that the ultraquick flash is of questionable physiological significance.

B. EXCITATION BY PERIPHERAL NERVES

Case and Buck (1963) found that decapitated photurids stimulated at the anterior end of the cord showed two discrete latency classes. Cord transit times of 110–145 msec made up one class and subtraction of the 70–85 msec latency for direct lantern stimulation indicated a cord conduction velocity of 10–20 cm per second. Some specimens gave a 90 msec conduction time (40–50 cm per second). Measurements of conduction time from the posterior cord to the lantern nerves made by Buck *et al.* (1963) revealed only 5–15 msec delay. Therefore, only a small part of the slow flash stimulus-response latency of 80 msec can be accounted for in terms of conduction time in the peripheral nerves. By covering the lantern with foil and observing the light emission through a single tiny hole, they demonstrated that tiny spots of lantern are capable of producing compound flashes. This suggested that the sequence of spontaneous excitation was similar in all areas of the lantern. They further showed that similar action potential bursts could be recorded from different sites in the lantern.

Because the last abdominal ganglion lies in the 6th abdominal (anterior lantern) segment, Hanson (1962) and Buck *et al.* (1963) were able to deganglionate the 7th abdominal (posterior lantern) segment by severing its nerve supply from the ganglion. They demonstrated that 21–24 hours after nerve transection, animals lost the ability to produce slow, but not quick flashes. From this, they inferred a loss of the neural effector junction and deduced that the quick flash was generated beyond the neural link. The ability to produce a quick flash in denervated lanterns was used by Hanson (1962) and Smalley (1965) to show that the photocytes were still functional after certain treatments that abolished spontaneous flashing.

Buck *et al.* (1963) have suggested a peripheral excitation pathway accounting for the observed latency classes in the adult organ. Noting that the nerves terminate between the tracheal end cell and the tracheolar cells, they adopt the term "end organ," first used by Kluss (1958) to indicate the coupling agent of unknown function between the peripheral nerves and the photocyte. The end organ, then, is composed of nerve endings, tracheolar cells and tracheal end cell. The excitation pathway involves peripheral nerve, end organ and photocyte sequentially. The overall response latency must involve a delay of peripheral nerve conduction, a delay of nerve ending–end organ interaction, a delay involving activation of the end organ and a delay of excitation of the photocyte. From the observed similarities in contour, facilitation, summation, fatigue and general neuroeffector response of the flashes, they feel that both slow and quick flashes are the same response at the photocyte level. They suggest that the response thresholds are in the order: photocytes > end organ > peripheral nerve. Stimulation via peripheral nerve, results in the slow flash; stimulation of the end organ can result in the quick flash; and direct stimulation of the photocytes may produce an ultraquick flash.

The different temperature coefficients found for the latencies of slow and quick flash and the loss of the slow flash upon denervation support their contention that the slow flash is neurally mediated while the quick flash is not. All things considered, however, the latencies still seem surprisingly long. Buck *et al.* (1963) suggest that 50–60 msec seems too long simply for pure conduction delay from nerve ends to end organ activation. The end organ-photocyte link of about 18 msec seems excessive also, but the mechanisms involved in this coupling process have neither been identified nor measured.

Because the nerves do not terminate on the photocytes in *Photuris,* it is possible that these cells are not excited by depolarization. This might explain the difficulty in producing an ultraquick flash. On this basis the end organ would trigger the photocytes by chemical means.

C. CENTRAL ASPECTS OF NEURAL CONTROL IN THE ADULT

While the peripheral neuroeffector machinery of the lantern may explain the gross characteristics of the adult flash, it is unlikely to account for the enormous variation in flash form found in many free flying males. Buck and Case (1961) have observed that the four-peaked twinkling flash of the male of a particular *Photuris* species may degenerate into single flashes under laboratory stimulation. Seliger *et al.* (1964) confirm this observation in Jamaican fireflies and also demonstrate that complex flashes can be emitted in the laboratory males which normally produce simple flashes in the field.

There is no doubt that the brain initiates the flash. Numerous observers (Verworn, 1892; Lund, 1911; Buck and Case, 1961) have confirmed that spontaneous flashing occurs only in individuals with intact brain-lantern connections. Case and Buck (1963) found that the stimulus threshold increases as one proceeds from the abdomen to the thorax, but drops again in the head, suggesting that the site of initiation resides here.

Granted that the central neural burst acts to trigger the flash, does it control other aspects of the flash as well? Recording with electrodes in the lantern, Case and Buck (1963) could see in *Photuris* a clear correspondence between gross volley structure and gross flash form and they demonstrated that single and double volleys correspondingly produced single and double flashes (Figs 10 and 11). The volley acted much like a trigger because flash duration appeared to be independent of volley duration. Flash triggering volleys have also been recorded in *L. lusitanica* (Buonamici and Magni, 1967). In this species, however, no obvious relationship between volley size and

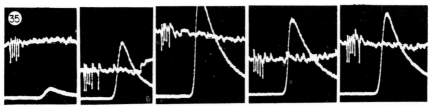

Fig. 10. Lantern potentials and spontaneous flashes in an intact Woods Hole *Photuris*, female. Action potential volley from anterior segment of lantern (upper trace) and subsequent spontaneous flash below. Oscilloscope frames equal 220, 220, 270, 270 and 270 msec respectively. (From Case and Buck, 1963.)

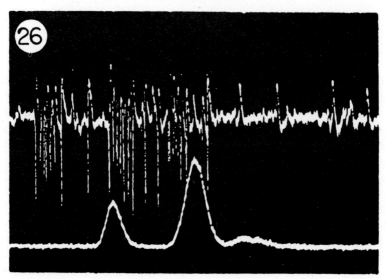

Fig. 11. Multiple action potential volleys and double flash in an intact Woods Hole *Photuris,* male. Oscilloscope traces as in Fig. 10. Oscilloscope frame equals 410 msec. (From Case and Buck, 1963.)

duration and response amplitude and contour were apparent. Aside from duration, Case and Buck (1963) could find little qualitative agreement between nerve volleys recorded from the ventral nerve cord and the subsequent volleys which appeared in the lantern. The time from first spike to initial flash rise was found to be 70–90 msec which correlates well with the latency of direct lantern stimulation. Volley duration and volley spike frequency could individually account for approximately 25% of the modulation of flash intensity. As volley frequency rose, spike frequency also rose and, combined, they could account for one-third of the overall influence on flash intensity. Isolated spikes outside of photic volleys appeared to have no effect on flash intensity, not even a priming function, as Buck and Case originally suggested.

Compound flashes are possibly dependent on central neural involvement. Case and Buck (1963) propose three mechanisms by which these flashes may be generated: (1) Sub-peaks are due to a series of central volleys; (2) The sub-peaks are due to photocyte populations responding at different latencies to a single volley; (3) Different neural conduction patterns exist, which result in asynchronous excitation of different photocyte populations (Buck,

1955). It would appear from this that much of the intensity modulation of the flash may reside in peripheral neural mechanisms outside of the flash generation centers.

A diurnal rhythm of C.N.S. excitability has long been suspected from the differences in frequency of spontaneous flashing with the time of day under uniform light conditions (Buck, 1937a). Peripheral modulation was first suggested by Case and Buck (1963), who found that the brain can not only play an important role in determining the animal's general excitability, but can also influence the excitatory state of the lantern as well. They found that intact animals, which are refractory to electrical stimulation, produce weak flashes with high thresholds. They can, however, be aroused by handling. Particularly important, they observed that the threshold of aroused animals remains low after decapitation or cord section. Refractory animals not aroused prior to decapitation stay in the refractory state when stimulated directly. From this, it would appear that the C.N.S. can determine the state of responsiveness of the entire excitation pathway. This is further suggested by the observation that refractory individuals, upon handling, often show an arousal sequence involving first a dim glow over the lantern, which brightens in blotchy spots, followed by a bright flash, then total extinction, then spontaneous flashing. This sequence suggests some priming process which must occur prior to sustained flashing.

Many observers have reported that bright light inhibits flashing in adult fireflies (Harvey, 1952). Case and Buck (1963) showed that electrical stimulation of the optic nerve of a spontaneously flashing specimen of *Photuris versicolor* could inhibit activity, which resumed soon after stimulation was stopped (Fig. 12). Localization of the electrode and stimulus intensity were critical, however. At low stimulus intensities flashing occasionally was inhibited but then resumed, even during stimulation. They conclude that in cases where activity was stopped, the stimulus is probably mimicking the effect of light which acts by direct central inhibition from the optic tract.

Case and Trinkle (1968) have produced flash inhibition by electrical or light stimulation of the eye of *Photuris missouriensis.* In order to suppress an expected flash the inhibitory stimulus had to be delivered no less than 80 msec earlier and if given more than 160 msec earlier, total inhibition always resulted. Light-dark transitions seemed to be more important than total light used, short artificial flashes being as inhibitory as brighter, longer flashes. Usually the lantern eventually escaped the inhibition and returned to its preinhibitory frequency even

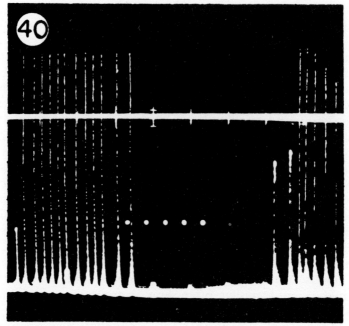

Fig. 12. Inhibition of spontaneous flashing during stimulation of eye of an intact Woods Hole *Photuris* male. The five dots denote stimuli to eye, 15 msec duration, 10 volts, 1 per second frequency. Oscilloscope frame equals 18 seconds. (From Case and Buck, 1963.)

under illumination, which suggested to the authors that light does not permanently affect the flash pacemaker mechanism.

Magni (1967) has taken advantage of the inhibitory effect of light to considerably amplify the role of central and peripheral modulation in flashing activity. At night *L. lusitanica* routinely produces 300 msec spontaneous flashes at frequencies of 60–80 per minute. Frequency and intensity of flashing can be inhibited by bright light and accelerated by dim light. Recording from the lantern revealed that when illumination was suddenly increased cessation of flashing occurred while neural volleys continued to be produced at a somewhat reduced rate. Upon cessation of the inhibitory light stimulus, flashing resumed at an accelerated rate that eventually returned to its original frequency and intensity. It is important to note here that separation of neural volleys and flash responses has been achieved. This strongly suggests that the inhibition occurs not only in the flash triggering center located in the brain, but acts peripherally as well. Illumination of the eye also inhibited both the peripheral and reflex flashes generated by direct, electrical stimulation of the lantern.

Artificially driven flashing by decapitated adults could be inhibited by stimulation of the ventral nerve cord. Stimulation of the cord at 200–500 per second increased flash intensity which returned to normal seconds after cessation of the stimulus to the cord. Voltages 4–5 times as high were required to induce a flash response by stimulating the cord. Low voltage cord stimulation of decapitated males at 20–50 per second produced total inhibition of flashing elicited by lantern electrodes. The animals could be induced to flash normally upon cessation of the inhibitory stimulus. Transection of the cord abolished the modulation, proving that it was neurally conducted and not due to electrical spread to the lantern.

Light apparently acts to influence the excitability of the animal for flashing centrally by affecting the flash generating center in the brain and peripherally by directly affecting the lantern. This peripheral inhibitory effect has been claimed to be humorally mediated by release of an inhibitory substance from the male gonads located in the 5th abdominal segment (Brunelli *et al.,* 1968a, 1968b). Gonadectomized males flash with much reduced variability in both frequency and intensity. However, the inhibitory effects of illumination are significantly reduced, but not abolished, by gonadectomy.

The experimental setup devised by Brunelli *et al.* (1968a, 1968b) to demonstrate and test these effects was ingenious (Fig. 13). Two males from which the tip of the terminal abdominal segment had been removed were joined across their abdominal tips by a saline bridge. While animal A was illuminated, the eyes of animal B were covered by opaque paste. Flash responses of each male were recorded separately by light pipes over the lanterns. Illumination of the eyes of animal A blocked his flashing entirely and significantly slowed the flash rate and intensity of animal B with a latency of 8–15 seconds when they were joined by the saline bridge. Inhibition of animal B could still be obtained after cord transection of animal A posterior to the 5th abdominal segment. It persisted even after removal of the entire lantern of animal A. Inhibition by illumination was abolished by cord transection of A anterior to the 5th segment, or by removal of both gonads without cord section.

Brunelli *et al.* (1968b) showed that the peripheral flash induced by direct lantern stimulation was no longer depressed in illuminated animals suffering bilateral gonadectomy. The longer latency reflex response was inhibited, however. Also, it was no longer possible to

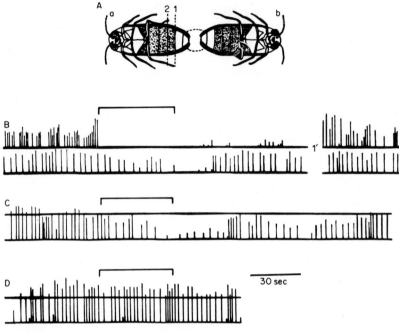

Fig. 13. Effect of illumination of firefly A on flashing of both fireflies A and B, before and after ablation of the lantern and of the 5th abdominal segment of firefly A. Throughout the experiment the two animals were kept in contact through a saline bridge. A: diagram of the experiment: the dotted lines 1 and 2 indicate the level of the sections performed to ablate the lantern and the 5th abdominal segment respectively. B: Record of the flashing of fireflies A and B before, during and after illumination of the intact firefly A with 100 Lux. The record taken after the break of 1 minute shows recovery of spontaneous activity of firefly A. C: same as A, but after the ablation of the lantern of firefly A, as shown in A by the dotted line 1. D: same as B, but after ablation of the 5th abdominal segment of firefly A (dotted line 2 in A). (From Brunelli *et al.,* 1968.)

inhibit directly induced flashes by low frequency cord stimulation of gonadectomized animals. In fact, this stimulation produced enhancement instead, which high frequency stimulation had alone been able to accomplish. Apparently in this case low frequency stimulation had produced both depression and enhancement in animals containing gonads but the inhibition had been stronger. The authors conclude that two separate systems must serve depression and facilitation of flash activity. They suggest that the inhibitory substance neurally released by the gonads affects flashing by acting at the neuro-effector junctions or at some other point in the lantern effector system. The inhibition which survives gonadectomy was conclusively shown to be central in origin and is most probably associated with reduction of central triggering volley output.

A neurally modulated inhibitory system, mediated through the release of a humoral agent, is apparently not a universal characteristic of control in fireflies because Case and Trinkle (1968) found no evidence for it in *Photuris missouriensis*. Flashes electrically driven in the lantern could not be inhibited by illumination of the eye, except when they were augmented by central activity. When the spontaneous flashes were inhibited by illumination of the eye the directly stimulated flashes were less intense, suggesting that the voluntary component had been previously facilitory. It was not found possible to transfer inhibition from one firefly to another using the experimental arrangement of Brunelli *et al.* (1968a). Noradrenaline injected into one *P. missouriensis* required a minimum transfer time of 68 sec to be detected in the second animal. This is in contrast to the 8–15 sec transfer time observed for the inhibitory humoral agent in *Luciola*. Case and Trinkle suggest that this transfer time in *Luciola* is possibly too short to be accounted for by diffusion of some humoral agent alone.

D. NEUROEFFECTOR CHARACTERISTICS OF THE LARVAL LANTERN

Although the larval lantern produces a spontaneous glow lasting for seconds it still shows many characteristics of the typical neuroeffector. Chang (1956) demonstrated that the isolated lantern can produce a response when stimulated with single shocks. Repetitive stimulation resulted in treppe and higher rates produced summation. Buck and Case (1961) demonstrated facilitation to paired shocks with maximal effect of the second shock coming 50–60 msec after the first. This facilitation lasts up to 500 msec compared to the 50 msec found in the adult. Stimulus-response latency was found to be a staggering 800 msec at 22°C. The temperature-latency relationship appeared to be linear between 14 and 31°C, compared to the curved relationship in the adult. The Q_{10} is 3.0 compared with 2.0 in the corresponding range for the adult. These observations all suggest fundamentally different neuroeffector mechanisms in the larva and adult. This is not too surprising when the differences in structure, such as the absence of tracheal end cells and close nerve-photocyte association in the larva, are considered. Buck and Case (1961) suggest that the tracheal end cells or end organs may be associated with the ability to quench luminescence rapidly. However, they do not accept that glowing in the adult necessarily means end cell inactivation.

At high stimulation intensities the larval lantern is capable of producing a flash of 1–3 msec response latency and a duration as low as 150 msec. While Buck *et al.* (1963) consider this response to be as dubious as the adult ultraquick flash, they showed that it possesses some effector characteristics. It facilitates and follows frequencies up to 10 per second in contrast to the larval glow shown by Chang (1956) to be unable to follow a stimulation frequency of more than 1 per second. This larval flash latency is temperature-independent. Three days following nerve transection the lanterns lose the ability to produce a slow, stimulated glow of 800 msec latency, but not the flash of 1–3 msec latency. The explanation suggested is that the ultra-quick flash of larvae represents direct stimulation of the photocytes as it may in the adult.

V. PHARMACOLOGY OF THE LANTERN

Acetylcholine has been found to be ineffective in inducing lumine-scence in the firefly. Case and Buck (1963) showed that the anti-cholinesterase, eserine, generates scintillation when injected into the animal at concentrations of 10^{-4} to 10^{-3} M. This scintillation is accompanied by uncoordinated neural bombardment of the lantern. Lanterns freshly deganglionated by transection of peripheral nerves will no longer scintillate. These observations point to the ganglia as the site of action of eserine in scintillation induction.

Smalley (1965) confirmed the original observation of Kastle and McDermott (1910) and Emerson and Emerson (1941) that adrenaline induces luminescence when injected into the adult. She demonstrated that the adult lantern behaves in closely analogous fashion to the vertebrate sympathetic system. Both adrenaline and noradrenaline were found to act at a threshold of approximately 10^{-4} M even on posterior segments denervated by nerve transection 48 hours pre-viously. She proposed that amphetamine acts indirectly by liberating transmitter from the lantern nerve endings because it failed to act in denervated portions of the lantern. Borowitz and Kennedy (1968) have shown that tyramine acts directly to induce luminescence in the same manner as noradrenaline in *Photinus pyralis* which is contrary to its indirect action in the vertebrate.

Taking advantage of the observation of Buck *et al.* (1963) that the slow flash is neurally mediated while the quick flash is generated farther down the neuroeffector chain, Smalley was able to detect effects of drugs acting at the nerve ends. Amphetamine and yohimbine, presumably by draining the terminals of transmitter, were found to block the slow flash. Reserpine, known to cause the release and destruction of adrenergic transmitters in vertebrates, blocked both slow and quick flashes when injected 36 hours previously. The quick flash could be restored immediately after injection of adrenaline, noradrenaline or extracts of corpus cardiacum. The restoring effect of these substances was thought to be due to stimulation of glycogen breakdown which was inhibited by reserpine. These observations are summarized in Table II.

Table II

Types of response to electrical stimulation after pretreatment with various compounds

Compound	No. of animals	No. of fireflies showing each type of response		
		No response	Slow and quick flash	Quick flash only
Saline	12	3	9	0
Norepinephrine	5	0	5	0
Amphetamine	12	2	0	10
Yohimbine	6	0	1	5
Reserpine	12	10	0	2
Reserpine plus 10^{-2} M norepinephrine or epinephrine	7	1	0	6
10^{-5} M norepinephrine	3	3	0	0
Corpus cardiacum extract	5	0	0	5

(From Smalley, 1965.)

By testing extracts of anterior and posterior portions of fireflies on rat uterus and colon Smalley demonstrated that adrenergic activity existed in both portions. The effect was not particularly straightforward because fresh extracts, instead of depressing acetylcholine induced contractions of uterus or colon, actually enhanced them. This effect was replaced by the typical inhibitory response after repeated exposures of the test muscles. The posterior abdomens were found to contain a higher concentration of inhibitory (adrenergic) substances than the anterior part. The uterus-colon activity ratios were intermediate between adrenaline and noradrenaline. The widespread occurrence of phenolic compounds related to adrenaline such as dopamine, particularly in the cuticle of insects, make these tests of adrenergic activity even more equivocal. Only the use of extracts obtained from isolated photogenic tissue holds any hope of eliminating the objection of unwanted phenolic analogs which show adrenergic activity.

Carlson (1968a) showed that adrenaline and noradrenaline also induced luminescence in the extirpated larval lantern immersed in oxygenated saline. A typical effect is shown in Fig. 14. A wide

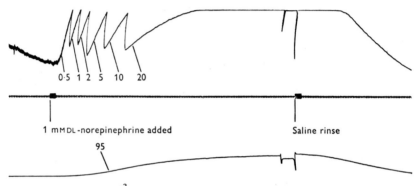

Fig. 14. Effect of 10^{-3} M noradrenaline on light response of extirpated larval lantern. Top and bottom traces, photomultiplier output; numbers refer to gain settings in mV/cm. Middle trace, time and event marker, 1 mark/sec. Sharp intensity drop on bottom trace indicates that noradrenaline-containing solution was removed. (From Carlson, 1968a.)

variety of catecholamines and phenols were tested (Carlson, 1968a, 1968b) and their potencies are summarized in Table III. Analysis of the results suggests the following general conclusions:

(1) The para hydroxyl group has a great effect on the potency of the drug, while the meta hydroxyl appears to be inhibitory (compare synephrine, adrenaline and L-phenylephrine).

(2) A methoxy group in the meta position does not reduce as much as an hydroxyl group (compare metanephrine with adrenaline and normetanephrine with noradrenaline).

(3) The terminal methyl group on the amino end appears to enhance potency (compare synephrine with octapamine and adrenaline with noradrenaline).

Table III

Drug	Relative Equivalent Concentration*	Lanterns tested
DL-Synephrine	1.00†	—
DL-Octapamine	6.19†	17
DL-Metanephrine	9.87†	18
DL-Normetanephrine	10.26†	15
DL-Adrenaline	43.66‡	14
DL-Noradrenaline	65.5	—
Dopamine	178.0‡	15
Tyramine	218.3‡	12
L-Phenylephrine	250.93†	15
DL-Isoproterenol	4093.7‡	8

* Concentration $\times 10^{-3}$ M at which drug will produce an effect equivalent to 10^{-3} M synephrine.
† Potency assessed by measuring maximum rate of intensity rise.
‡ Potency assessed by measuring maximum intensity of luminescence.
DL-Noradrenaline used in both types of potency measurement.
(Modified from Carlson, 1968a, 1968b.)

(4) An hydroxyl group on the β-carbon appears to increase potency (compare noradrenaline with dopamine and octapamine with tyramine).

(5) Substitution of an isopropyl group for a methyl group on the amino end drastically reduces potency (compare adrenaline with isoproterenol). In summary, nearly all parts of the molecule have effect on the potency of its luminescence inducing action.

At present the true chemical structure of the larval, and presumably adult, transmitter can only be speculated upon. Synephrine, acting at a threshold of 10^{-6} M, is the most potent drug so far tested. It, along with all the others shown in Table III, acts in typical fashion showing a sigmoidal dose-response curve when the maximum intensity attained, or the maximum rate of intensity rise, is measured against drug concentration. This is in distinction to amphetamine which clearly acts indirectly on the nerve ends to release transmitter. Amphetamine produces a slowly rising glow which resists extinction. It acts with greatly reduced effect in larvae denervated or reserpinized 48 hours previously.

Is the true transmitter a monophenol resembling synephrine? Aside from the observations on potency of luminescence induction no other evidence has been developed to support or disprove this hypothesis. A monophenolic transmitter may be necessary so that it can be destroyed by a tyrosinase-type enzyme from the blood. There may be some relationship between a monophenolic transmitter and the need to maintain a transparent cuticle. The elucidation of this problem awaits chemical isolation studies and investigation of enzymatic breakdown of mono and dephenolic substrates.

It has further been shown that noradrenaline induces luminescence not only in saline but also in Na^+ and K^+ free solutions such as isotonic sucrose and choline chloride. This would suggest that the drug does not act by controlling the movement of cations through the cell membrane. High K^+ saline induces an intense luminescence but fails to act on denervated or reserpinized lanterns suggesting that it acts to release transmitter by depolarization of the nerve terminals.

10^{-3}M KCN rapidly and reversibly extinguishes the glowing lantern even when introduced with a high concentration of synephrine. It presumably acts to quench ATP production and thereby limit the amount of active complex capable of reacting with oxygen to produce light. Is a pool of ATP present in the lantern to carry the luminescence induced by neural stimulation? Examination has revealed that the time required to extinguish the lantern is directly proportional to luminescence intensity or synephrine concentration (Fig. 15), not

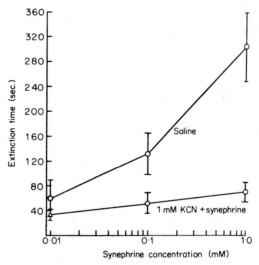

Fig. 15. Effect of synephrine concentration on the luminescence extinction time produced by 10^{-3}M KCN plus synephrine and by saline. Vertical lines through the points equal to standard error of the mean to 95% level of confidence. (From Carlson, 1968b.)

inversely proportional as expected for a large ATP pool. If a large ATP pool were available, one might expect KCN to require a significantly longer period to extinguish the lantern at low luminescence intensities because the ATP pool would not be so rapidly exhausted. It appears then that synephrine stimulates the production of ATP at a rate proportional to its concentration. Sutherland and Cori (1951) have shown that catecholamines stimulate ATP production in liver by increasing phosphorylase activity and glycogenolysis. Amphetamine may also stimulate ATP production in larval lanterns. Reserpinized lanterns, tested with noradrenaline before and after amphetamine treatment, showed greatly enhanced responses after soaking in amphetamine. In eight lanterns tested, the post-amphetamine responses to noradrenaline averaged 426% greater than those prior to amphetamine treatment (Carlson, 1968a).

Rall and Sutherland (1962) have demonstrated that catecholamines act with adenyl cyclase to produce cyclic adenosine-3',5'-phosphate and pyrophosphate in the vertebrate. This may serve as a possible model of synephrine and catecholamine action in larval luminescence induction. Pyrophosphate, shown to be active in *in vitro* luminescence induction by McElroy (1951), could be liberated by a mechanism such as this. It would then be generated directly by the transmitter, rather than being released through a number of intermediate chemical steps as proposed for acetylcholine by McElroy and Hastings (1955).

VI. OTHER LUMINESCENT RESPONSES

A. SCINTILLATION

Scintillation, the rapid, uncoordinated flashing of hundreds of individual elements, can be produced by a wide variety of agents such as spider venom (Wood, 1939), cyanide (Buck, 1948), and eserine (Case and Buck, 1963). Even in the untreated animal, after a period of quiescence scintillation and uncoordinated, splotchy bursts of light often proceed a coordinated flash. This in itself suggests that the normal neural burst must trigger the elements in the lantern in near simultaneous fashion, in order to produce a burst of light of 200 msec duration.

While eserine and other agents, such as spider venom, induce scintillation at the ganglion level, Carlson (1967) has shown that saline solutions high in K^+ can induce scintillation for periods of up to one hour when applied directly to the photogenic tissue exposed by cuticle removal. This scintillation is not extinguished by deganglionation and no random nerve activity can be detected in the scintillating lantern. Functional nerve endings are apparently required for the scintillation response. Scintillation cannot be induced in portions of the lantern in which nerve degeneration has been produced, while adjacent innervated portions scintillate readily. The observation that only animals capable of vigorous flashing can be induced to scintillate further supports the hypothesis that functional nerve ends are required. Table IV summarizes the effect of various salines applied to the exposed photogenic tissue. These observations suggest the following general conclusions:

(1) K^+ ions are mandatory for induction of this scintillation response. Ca^{2+} ions are not essential but greatly facilitate the response (see lines 1–4).

(2) Mg^{2+} ions cannot replace Ca^{2+} ions in inducing scintillation (see line 5 vs. lines 1 and 2).

(3) As the ratio of Na^+ to K^+ is increased, the scintillation response shifts to a flash response. Na^+ appears to be a requirement for the flash response (see lines 10–13 and Fig. 16).

An assessment of the mechanism of action of scintillation inducing salines is difficult due to the large number of intervening events in the neuroeffector chain. The high K^+ level may induce scintillation by depolarizing either or both the coupling cell or the nerve terminals. If the latter are depolarized this might increase transmitter release in terms of quantity or frequency. Depolarization of the photocytes as well cannot be ruled out but if that occurred to any extent one might expect a permanent glow to result.

B. PSEUDOFLASH

The pseudoflash, first reported by Snell (1932), is a flash of over one second duration produced by allowing the in-diffusion of oxygen to an adult or larval lantern made anoxic in nitrogen (Figs 1 and 2). As anoxia proceeds in a spontaneously flashing adult, the flashes increase in duration and drop in intensity. A low level glow called the hypoxic glow begins to develop which slowly reaches a maximum and then declines. If oxygen is readmitted at any time during the hypoxic glow a pseudoflash results. This flash is clearly due to the rapid oxidation of built-up active complex but its duration, as shown by Hastings and Buck (1956), is very constant. Carlson (1961) has shown that neural activity is a requisite to the build-up of active complex during

Table IV

Effects of ions on light response of exposed organ of *Photuris* males

No.	KCl (M)	NaCl (M)	CaCl$_2$ (M)	MgCl$_2$ (M)	Sucrose (M)	Other (0.16M)	Light response* Scint†	None‡	Flash§
1	0.16	0	0.002	0	0	0	113	60	0
2	0.16	0	0.006	0	0	0	7	1	0
3	0.16	0	0	0	0	0	10	6	0
4	0	0	0.105	0	0	0	0	4	0
5	0.16	0	0	0.006	0	0	0	4	0
6	0.12	0	0.006	0	0.062	0	6	1	0
7	0.10	0	0.006	0	0.102	0	4	0	0
8	0.08	0	0.006	0	0.142	0	3	1	0
9	0.06	0	0.006	0	0.182	0	0	2	0
10	0.12	0.04	0.006	0	0	0	4	0	0
11	0.08	0.08	0.006	0	0	0	4	2	2
12	0.04	0.12	0.006	0	0	0	0	0	2
13	0	0.16	0.006	0	0	0	0	0	10
14	0	0	0.006	0	0	KI	5	0	0
15	0	0	0.006	0	0	KBr	2	0	0

* All test solutions were between pH 6.5 and 6.9. Control scintillation lasting more than 5 minutes.
† Scintillation response lasting over 5 minutes.
‡ Scintillation duration less than 5 minutes.
§ Exposed area flashes with remainder of lantern.
(From Carlson, 1967.)

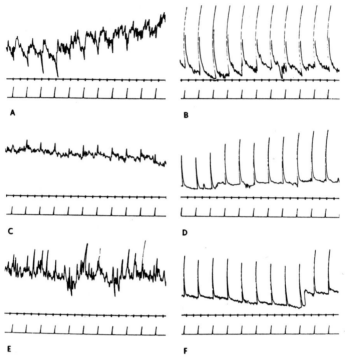

Fig. 16. Three successive alternating periods of scintillation and flashing induced in a single exposed right side of anterior photogenic segment of *Photuris* male. Nerves to posterior lantern segment transected and left side of anterior segment masked with black tape. Upper trace is photomultiplier output. Middle trace is time base, 1 mark/sec. Lower trace is stimulus of 15 volts and 20 msec duration. Stimulating electrodes on segment anterior to front lantern segment. Each pair of records is at same amplification. (A) Scintillation 5 min after application of "scintillation solution" (composition given on line 1, Table IV). In this and all subsequent records figures in parentheses indicate total time elapsed in seconds between application of solution and beginning of record (300). (B) Flash response 17 sec after being perfused with a solution of composition shown on line 13, Table IV (380). (C) Scintillation 30 sec after perfusion with solution as in record (A) (488). (D) Flash response 50 sec after perfusion with solution as in record (B) (600). (E) Scintillation 81 sec after perfusion with solution as in record (A) (772). (F) Flash response 62 sec after perfusion with solution as in record (B) (932). (From Carlson, 1967.)

anoxia. This makes less tenable any suggestion that anoxia is directly responsible for the production of active complex by the removal of some active quenching process. While the larva produces a similar pseudoflash no preceding hypoxic glow can be produced (Carlson, 1965). This suggests that this glow may be a function of the coupling cells, perhaps the tracheal end cell.

C. GLOWS

A large variety of treatments produce steady glows from the adult lantern. A number of metabolic inhibitors in low concentration, such as KCN, induce glowing (Kastle and McDermott, 1910) and the adult often glows after death. This suggests that some active quenching process maintains the lantern in the darkened state. Because of its apparently high metabolic activity, suggested by its dense concentration of mitochondria, the end cell might be the site of this effect (Buck and Case, 1961). This is particularly significant in that larval lanterns are quite resistant to glow production by these means (Carlson, 1965). It is possible that the photocyte flash mechanism is in some way energy dependent and destruction of this energy source, combined with nerve activity or transmitter release, results in a glow rather than a flash or scintillation. It should be mentioned here that high concentrations of KCN of the order of 10^{-3} M actually quench luminescence indicating that the light reaction itself requires energy but is less susceptible to inhibition.

VII. DEVELOPMENT OF THE ADULT LANTERN

The adult lantern forms during the pupal period of 16–18 days (Hess, 1922), and although the larval lanterns persist during this period, extirpation experiments by Harvey and Hall (1929) have confirmed that they do not contribute to the formation of the adult lantern. Because the various structures and processes which are involved in flashing would not be expected to develop simultaneously, examination of the changing luminescent capability of the pupa could reveal important information on underlying neural mechanisms. The early workers (Williams, 1916; Dahlgren, 1917; Hess, 1922) who investigated the histological development of the adult lantern provide very little information on flash genesis. This is to some extent understandable because most of the structural organization observable with the light microscope is completed during the early pupal period while the glow changes into a flash during the 24 hours following eclosion.

C. Kaars, personal communication, has recently studied the development of flashing ability in the adult lantern of *Photuris* pupae and his observations comprise the following account.

(1) A low intensity glow, readily noticeable to the dark adapted eye at 12 inches, appears over the entire pupa 3–5 days after pupation, confirming the observations of Williams (1916). This glow persists

for the entire pupal period but becomes harder to see as the epidermal pigmentation becomes darker. This pupal glow does not change noticeably in intensity when the animal is flushed with oxygen or mechanically stimulated.

(2) Luminescence first appears in the developing adult lantern on the 12th or 13th day of pupation. It is normally very dim but can be readily enhanced by mechanical stimulation of the head and has a time course similar to that of the larval organs (Fig. 17). This luminescence does not appear to change spontaneously. When pupae were monitored overnight by photomultiplier no fluctuations of intensity were found.

(3) Neural bursts of 4–12 action potentials accompanied the prolonged glows of newly emerged adults (Fig. 18). The frequency of these bursts was high as the lantern brightened and fell off during the sustained glow which followed, compare Fig. 18b and c. These observations suggest that the central neural mechanisms are in functioning order at this point and the inability to flash is due to peripheral mechanisms in the lantern itself.

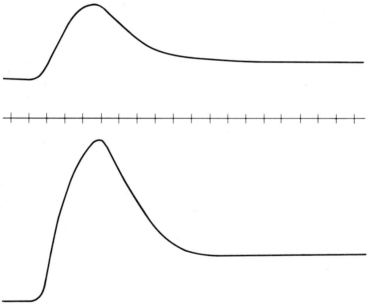

Fig. 17. Response of adult lantern (upper trace) and larval lantern (lower trace) of a pupa 48 hours prior to eclosion in response to mechanical stimulation. Light was recorded by separate light pipes positioned over each lantern. Time marks (center trace) at 1 second intervals. (From Kaars, unpublished.)

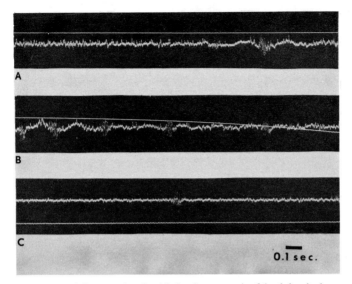

Fig. 18. Neural activity associated with luminescence in 6th abdominal segment of a pupa just prior to eclosion. A. and B. Continuous records of neural activity (lower trace) recorded by electrodes in the lantern and light intensity (upper trace) which increases with downward deflection of trace. Perturbations of lower trace due to movement of animal in response to mechanical stimulation. C. Neural activity (upper trace) during sustained glow (lower trace) 45 seconds after the initiation of the glow seen in B. (From Kaars, unpublished.)

(4) This lag in peripheral mechanism is demonstrated in Fig. 19 which shows that a pseudoflash induced 24 hours prior to eclosion is considerably slower than the typical one induced 30 hours later in the same animal. Development of the tracheal system may explain the reduction of pseudoflash duration during this period.

(5) First flashes were observed 24–36 hours after eclosion (Fig. 20). The flashes began in small, homogeneous regions of the anterior lantern segment. These regions did not seem to necessarily represent fields of photocytes innervated by the same nerve (Hanson, 1962). The flashes were greater than 600 msec duration which was due to slow rise times as well as slow decay times. These slow rise and decay times did not appear to be the result of lack of co-ordination of the various areas which had developed flashing ability.

VIII. THE NEUROEFFECTOR RESPONSE UNIT AND ITS CONTROL

The ultimate description of neural control of luminescence in the firefly hinges on the correct determination of the smallest functional

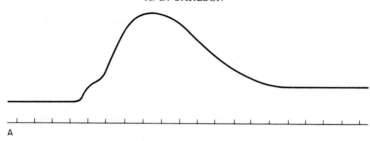

A

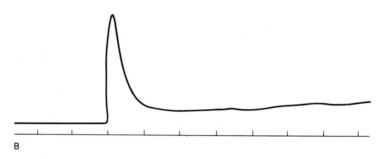

B

Fig. 19A. Pseudoflash induced in 6th abdominal segment by placing an anoxic pupa in a stream of oxygen, 24 hours prior to eclosion. Time marks on lower trace: 1 mark per second in both A and B.

Fig. 19B. Same as A but approximately 6 hours after eclosion. (From Kaars, unpublished.)

lantern neuroeffector unit. This can be defined as the smallest unit capable of producing a flash response with neural stimulation.

On the basis of visual examination of flashing and glowing lanterns, Buck (1966) considers the tracheal cylinder to be the smallest normal physiological unit. The cylinder is comprised of dorso-ventral tracheal trunk and associated nerve trunk along with tiers of end organs. These include groups of tracheal end cells surrounded in rosette fashion by the photocytes and tracheolar cells which penetrate between and within the photocytes. Buck bases this conclusion on the observation that this collection of effectors flash in near simultaneity, although a rapid ventrad spread of luminescence through the cylinder is sometimes observed. Secondly, partial flashing of a cylinder in a normal lantern is rare. Dimly glowing lanterns show rings of luminescence surrounding dark cores in the region of the tracheal

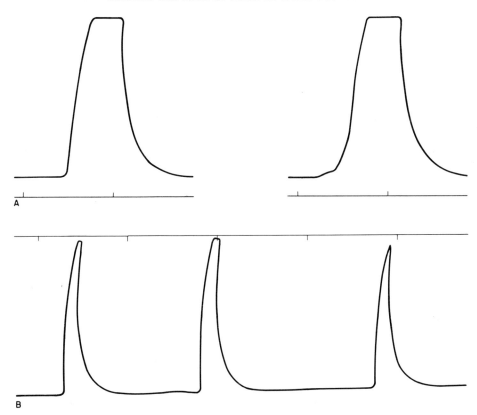

Fig. 20A. The 12th and 36th flashes produced by a newly emerged adult. Time marks on lower trace: 1 mark per second in both A and B.

Fig. 20B. Spontaneous flashes produced by a field collected adult male for comparison. (From Kaars, unpublished.)

trunks. The exact location of this luminescing material was difficult to assess, however, because of diffusion of light within the lantern tissue. If it was truly from the photocytes, it seemed most likely to be the differentiated zone on the periphery (Lund, 1911).

Buck (1966) also reports observing responses from small point units within the photocyte area which differ from the response of cylinder units in two ways. They usually produce a brighter flash which stands out of the duller glowing surround and their flash duration is considerably shorter, of the order of 20 msec. These point units are capable of the same range of activity as the cylinder unit. They appear to be single photocyte responses. Scintillation caused by a number of treatments can be composed of both types of responses but analysis of unit response activity is impossible via photomultiplier recording and extremely difficult by microscopic observation because of its rapid, unco-ordinated and random character.

Hanson *et al.* (in press) have considerably reduced the uncertainty of unit analysis by an examination of *Photuris* light responses using high resolution image intensification and cinematography. Comparison of image intensified photographs of glowing regions and histological sections of similar regions enabled them to localize the luminescent region to that associated with the photogenic granule-filled interior of the photocytes. They observed spots of light, called micro-sources, which appeared to be the smallest functional units of the flash. These microsources had identical characteristics in submaximal flashes, scintillation and in pre- and post-flash spots. They were normally circular and had mean diameters of 31μ but tended to grow and shrink during their lifetime. Their durations were variable, e.g. in submaximal flashes from 30–350 msec and were normally distributed about an average of 139 ± 3 msec. However, micro-sources that appeared later in the flash were significantly shorter in duration; they appeared in successive flashes, but were rarely extin-guished and regenerated during the same flash. Microsources were found to be significantly smaller than individual rosettes; in fact, they could be seen flashing within rosettes. These workers could only find glowing, never flashing, rosettes in their records and seriously doubt that cylinders, on this basis, could be the functional units.

If it is assumed that only the central region of a photocyte is actively luminous, the microsources were closely equivalent to 2 to 3 adjacent photocytes which measure about 21.5μ in luminous length and $6–10\mu$ in width. They suggest that the smallest functional flash unit is best considered equivalent to an end organ complex of tracheal end cell and associated photocytes, as suggested by Buck *et al.* (1963) and as pictured by Smith (1963). This is further streng-thened by the fact that the mass flash could be accounted for by the recruitment of microsources of variable but shorter duration. Their technique could not resolve units of the size reported by Buck (1966) with 20 msec flash duration, so the possibility still exists that smaller functional units do exist within the lantern.

In spite of this new information, one is still not able to distinguish between the two possibilities of neural control suggested by Smith

(1963). Neural excitation of the photocytes could be produced by (1) diffusion of transmitter through the end cell to its associated photocytes; or (2) depolarization of the tracheolar cells which may channel the excitation to the photocytes. Peterson and Buck (1968) point out that the discovery of end organ complexes in Asiatic fireflies is further proof that these structures are required in adult firefly flash production. Their discovery that the nerves in these lanterns terminate on the photocytes as well as between the tracheal end cell and tracheolar cell need not rule out coupling via the end organ complex as suggested by Buck *et al.* (1963).

That the difference between flashing and glowing forms is due to an end organ complex is further strengthened by the observations on pupae. At early stages, the larval and developing adult lanterns give identical responses, but this non-flashing pupal lantern receives adult-like neural bursts. This would suggest that central activity cannot be crucial in controlling the ability to produce a flash.

It is possible that some fundamental difference exists between larval and adult photocytes which can explain much of the difference in luminescent kinetics. The so-called flash responses induced in isolated larval lanterns by electrical stimulation are significantly longer than adult flashes. This is true of those obtained by Chang (1956) and Buck *et al.* (1963). Only with very high voltages could the latter investigators induce small flashes comparable in duration to the adult flash. McElroy (personal communication) has observed that larval lanterns are significantly lower than adult lanterns in the enzyme inorganic pyrophosphatase, which destroys pyrophosphate, the reactant presumed to link excitation with luminescence. This alone might explain why it is difficult to induce short duration flashes in larval lanterns.

It is possible that slow breakdown of released neural transmitter is at least partly responsible for the long larval glow. Following this reasoning, if the adult is to produce a flash it must develop a system which either rapidly destroys the transmitter or, failing that, acts in some way to reduce its duration of action on the effector cell. The tracheal end cell would be the best candidate for this latter role because it represents the fundamental cellular difference between larval and adult forms. If this suggestion were correct, short duration flashes might be more readily induced in directly stimulated larval lanterns suffering nerve degeneration from transection of peripheral nerves. This has not been attempted as yet.

IX. ANALYSIS OF BRAIN FUNCTION

The adult flash response provides an excellent tool for the analysis of brain function. It is relatively all-or-none, easily observed and recorded and it need not involve movement of the test animal which would alter the experimental parameters. Further, it is a relatively unambiguous executive act which signifies, under appropriate conditions, acceptance of a flash signal. A number of investigators have taken advantage of this response to study some brain phenomena, examples of which are synchronous flashing, signal and species discrimination and mimicry.

A. SYNCHRONOUS FLASHING

Buck (1938) has shown that the *Photinus pyralis* male is capable of synchronizing with other males. The most impressive displays, however, are produced by the oriental genus *Pteroptyx* and may involve hundreds of individuals flashing twice a second in synchrony. Buck and Buck (1968) demonstrated that the synchrony of *P. malaccae* was not an illusion as originally proposed. It was actually within ± 16 msec of perfect coincidence. They have shown that *P. malaccae* males are capable of a high order of regularity in their individual flashing rhythm. In one individual, with a mean flash cycle length of 557.3 msec, 95% of all the flash cycles fell within 5 msec of the mean period. They convincingly argue that the mechanism of synchronization must involve anticipatory or "sense of rhythm" synchrony. This implies that the animal, to attain synchrony, must initiate the neuroeffector events necessary to trigger its flash nearly a cycle before the flash actually occurs. *P. malaccae* was found to have lantern latencies to direct electrical stimulation of 55–80 msec and a presumed cord conduction time of at least 15 msec. This total of at least 70 msec response latency is much longer than the 16 msec synchrony variation and effectively rules out the possibility of entraining by direct contemporary visual stimuli from adjacent fireflies. This is an impressive display of timing ability for any insect central nervous system.

B. SIGNAL AND SPECIES DISCRIMINATION

The firefly flash is recognized as the mechanism of attraction of the sexes. McDermott (1914) was one of the first to quantitatively analyse male-female flash exchanges. Barber (1951) used these as a basis for taxonomic identification, which he termed physiological species. He showed that while sympatric species could be isolated by habitat and time or period of activity, the flash sequence was also a powerful isolating mechanism. Lloyd (1966) made a comprehensive study of the distribution and flash behavior of North American *Photinus* fireflies. He demonstrated that females could discriminate males of their own species on the basis of flash duration, or in the case of double flashing males, flash number or flash interval. His experimental approach included the laboratory testing of the female's ability to discriminate flashes of different durations. Captured females of sympatric species placed in close proximity were also judged on ability to attract free flying males in the field.

What, then, are the parameters of the male flash which may be of significant communicative importance to the female? A fairly comprehensive list would mention (a) color, (b) movement of stimulus, (c) altitude, (d) intensity, (e) background illumination, (f) pulse shape, (g) pulse duration, (h) pulse number, (i) pulse interval, and (j) phrase interval (period from the male flash of the male-female exchange to the male flash of the subsequent male-female exchange).

Flash color (Mast, 1912; Buck, 1937b), movement of stimulus (Mast, 1912), altitude (Lloyd, 1966) and flash intensity (Mast, 1912) appear to be parameters of no signal value to the female. Background illumination is important in influencing propensity to flash (Magni, 1967), but it is not important in discrimination of flash pattern. Pulse shape must play a significant role but has not as yet been sufficiently analysed. The other parameters, g–j, have all been shown to be of great influence on the ability of the female to recognize the male of the correct species (Lloyd, 1966). It should be noted that the effective parameters all involve the basic process of timing, shown to be a paramount ability of the firefly in synchrony studies.

Unpublished observations in my laboratory, on the female of a *Photinus* species found on Long Island which responds to a two flash phrase, reveal that discrimination involves mainly timing processes rather than any more complex analysis of signal content. In this situation the females have been phase shifted by 12 hours and male flashes are provided by a stimulator controlled pinlight bulb. As shown by Buck and Buck (1966) on *Photinus consanguineous,* the female answer is timed on the 2nd male flash while the first male flash is much less important. In the laboratory our female responded most readily (to greater than 90% of the signals) to a flash sequence of two 125 msec flashes, 1.2 seconds apart with a response coming

2.1 seconds after the onset of the initial test flash. The optimum phrase interval was found to be 5.2 seconds. Duration of the first male flash could be varied over wide limits. In this case, responses were obtained to double pulses in which the first pulse duration was increased up to 800 msec. The duration of the 2nd pulse could not be increased beyond 375 msec before responses failed. Responses could be obtained between flash intervals of 1.0 and 1.5 seconds with an optimum of 1.2 seconds.

Is the female susceptible to extra signals arriving during the male flash period, or does she systematically ignore extra information arriving during this time and responds as long as properly timed male flashes are present? It should be remembered that neural burst generation time occupies no more than 200 msec of the period following the 2nd male flash. This still leaves 600 msec unaccounted for during which the female is presumably still not committed to a flash answer. While interposition of an extra flash between the male flashes effectively cancelled female response, she did continue to answer upon addition of an extra flash within 500 msec after the 2nd male flash. This period after the 2nd male flash, then, appears to be essentially a non-analytical timing period. Even the effectiveness of the extra flash interposed between the two male flashes can be explained as interfering with the important timing of the male flash interval. Obviously further exploration in depth of signal parameters in this fashion for a number of firefly species should yield information on the processing ability of the insect brain. This behavioral approach, combined with studies of visual ability via electroretinograms, could help to sort out the essential features of the complex displays many species produce during flight.

C. AGGRESSIVE MIMICRY

Lloyd (1966) reported that *Photuris* females were capable of mimicking the response latencies of female *Photinus* species in their immediate area. By this behavior these large, voracious females were able to attract the *Photinus* males, capture and eat them. This report immediately suggests a substantially increased range of capability for the firefly brain beyond the machine-like, preset response to signals of species specific pattern. While the hypothesis of aggressive mimicry certainly adequately explains Lloyd's observations, it might be worthwhile to carry out a systematic laboratory study of this behavior. In my experience fireflies of the *Photuris* genus are extremely difficult behavioral subjects which flash erratically to many types of provocation. This in itself may suggest a wider response latitude which could encompass aggressive mimicry. Still, before one accepts the hypothesis of aggressive mimicry, with all that it implies, it might be worthwhile to determine whether the *Photuris* female's response latency can be changed by varying signal parameters in some systematic way. That this may be possible is suggested by the observation that response latency of the *Photinus* species females we studied was frequently lengthened by increasing the duration of the first stimulus (male) flash. Incorrect female response latencies may have a systematic physiological basis and may be triggered in rigid fashion by incorrect male signals which nevertheless are capable of initiating a female response.

REFERENCES

Alexander, R. S. (1943). Factors controlling firefly luminescence. *J. cell. comp. Physiol.* **22**, 51–71.

Barber, H. S. (1951). North American fireflies of the genus *Photuris. Smithson. misc. Collns* **117**, 1–58.

Beams, H. W. and Anderson, E. (1955). Light and electron microscope studies on the light organ of the firefly *(Photinus pyralis)*. *Biol. Bull. mar. biol. Lab., Woods Hole* **109**, 375–393.

Borowitz, J. L. and Kennedy, J. R. (1968). Actions of sympathomimetic amines on the isolated light organ of the firefly *Photinus pyralis. Arch. int. Pharmacodyn. Thér.* **171**, 81–92.

Brunelli, M., Buonamici, M. and Magni, F. (1968a). Mechanisms for photic inhibition of flashing in fireflies. *Archs ital. Biol.* **106**, 85–99.

Brunelli, M., Buonamici, M. and Magni, F. (1968b). Effects of castration on the inhibition of flashing in fireflies. *Archs ital. Biol.* **106**, 100–112.

Buck, J. B. (1937a). Studies on the firefly. I. The effects of light and other agents on flashing in *Photinus Pyralis,* with special reference to periodicity and diurnal rhythm. *Physiol. Zoöl.* **10**, 45–58.

Buck, J. B. (1937b). Studies on the firefly. II. The signal system and color vision in *Photinus pyralis. Physiol. Zoöl.* **10**, 412–419.

Buck, J. B. (1938). Synchronous rhythmic flashing of fireflies. *Q. Rev. Biol.* **13**, 301–314.

Buck, J. B. (1948). The anatomy and physiology of the light organ in fireflies. *Ann. N.Y. Acad. Sci.* **49**, 397–482.

Buck, J. (1955). Some reflections on the control of bioluminescence. In *The Luminescence of Biological Systems* (F. H. Johnson, ed.), pp. 323–333. A.A.A.S., Washington, D.C.

Buck, J. (1966). Unit activity in the firefly lantern. In *Bioluminescence in Progress* (F. H. Johnson and Y. Haneda, ed.), pp. 459–474. Princeton University Press, Princeton, New Jersey.

Buck, J. and Buck, E. M. (1966). Photic signalling in the firefly *Photinus consanguineus. Am. Zool.* **5**, 682.

Buck, J. and Buck, E. (1968). Mechanism of rhythmic synchronous flashing of fireflies. *Science, N.Y.* **159**, 1319–1327.

Buck, J. and Case, J. F. (1961). Control of flashing in fireflies. I. The lantern as a neuroeffector organ. *Biol. Bull. mar. biol. Lab., Woods Hole* **121**, 234–256.

Buck, J., Case, J. F. and Hanson, F. E. Jr. (1963). Control of flashing in fire-flies. III. Peripheral excitation. *Biol. Bull. mar. biol. Lab., Woods Hole* **125**, 251–269.

Buonamici, M. and Magni, F. (1967). Nervous control of flashing in the firefly *Luciola italica L. Archs ital. Biol.* **105**, 323–338.

Carlson, A. D. (1961). Effects of neural activity on the firefly pseudoflash. *Biol. Bull. mar. biol. Lab., Woods Hole* **121**, 265–276.

Carlson, A. D. (1965). Factors affecting firefly larval luminescence. *Biol. Bull. mar. biol. Lab., Woods Hole* **129**, 234–243.

Carlson, A. D. (1967). Induction of scintillation in the firefly. *J. Insect Physiol.* **13**, 1031–1038.

Carlson, A. D. (1968a). Effect of adrenergic drugs on the lantern of the larval *Photuris* firefly. *J. exp. Biol.* **48**, 381–387.

Carlson, A. D. (1968b). Effect of drugs on luminescence in larval fireflies. *J. exp. Biol.* **49**, 195–199.

Case, J. F. and Buck, J. (1963). Control of flashing in fireflies. II. Role of central nervous system. *Biol. Bull. mar. biol. Lab., Woods Hole* **125**, 234–250.

Case, J. and Trinkle, M. S. (1968). Light-inhibition of flashing in the firefly *Photuris missouriensis. Biol. Bull. mar. biol. Lab., Woods Hole* **135**, 476–485.

Chang, J. J. (1956). On the similarity of response of muscle tissue and of lampyrid light organs. *J. cell. comp. Physiol.* **47**, 489–492.

Chiarodo, A. J. (1968). The fine structure of neurons and nerve fibres in the thoracic ganglion of the blowfly, *Sarcophaga bullata. J. Insect Physiol.* **14**, 1169–1175.

Dahlgren, U. (1917). The production of light by animals. *J. Franklin Inst.* **183**, 79–94, 211–220, 323–348, 593–624.

Emerson, G. A. and Emerson, M. J. (1941). Mechanism of the effect of epinephrine on bioluminescence of the firefly. *Proc. Soc. exp. Biol. Med.* **48**, 700–703.

Hanson, F. E. Jr. (1962). Observation on the gross innervation of the firefly light organ. *J. Insect Physiol.* **8**, 105–111.

Hanson, F. E., Miller, J. and Reynolds, G. T. (In press) Subunit coordination of the firefly light organ.

Harvey, E. N. (1952). *Bioluminescence,* Academic Press, New York. 649 pp.

Harvey, E. N. and Hall, R. T. (1929). Will the adult firefly luminesce if its larval organs are entirely removed? *Science, N.Y.* **69**, 253–254.

Hastings, J. W. and Buck, J. (1956). The firefly pseudoflash in relation to photogenic control. *Biol. Bull. mar. biol. Lab., Woods Hole* **111**, 101–113.

Hess, A. (1958). The fine structure of nerve cells and fibers, neuroglia, and sheaths of the ganglion chain in the cockroach *(Peri-planeta americana). J. biophys. biochem. Cytol.* **4**, 731–742.

Hess, W. N. (1922). Origin and development of the light-organs of *Photuris pennsylvanica* DeGeer. *J. Morph.* **36**, 245–277.

Kastle, J. H. and McDermott, F. A. (1910). Some observations on the production of light by the firefly. *Am. J. Physiol.* **27**, 122–151.

Kluss, B. C. (1958). Light and electron microscope observations on the photogenic organ of the firefly, *Photuris pennsylvanica,* with special reference to the innervation. *J. Morph.* **103**, 159–185.

Lloyd, J. E. (1965). Aggressive mimicry in *Photuris:* Firefly femmes fatales. *Science, N.Y.* **149**, 653–654.

Lloyd, J. E. (1966). Studies on the flash communication system in *Photinus* fireflies. *Misc. Publs Mus. Zool. Univ. Mich.* No. 130. 95 pp.

Lund, E. J. (1911). On the structure, physiology and use of photogenic organs, with special reference to the Lampyridae. *J. exp. Zool.* **11**, 415–467.

Magni, F. (1967). Central and peripheral mechanisms in the modulation of flashing in the firefly *Luciola italica* L. *Archs ital. Biol.* **105**, 339–360.

Mast, S. O. (1912). Behavior of fire-flies *(Photinus pyralis)?* with special reference to the problem of orientation. *J. Anim. Behav.* **2**, 256–272.

McDermott, F. A. (1911). Some further observations on the light-emission of American Lampyridae: the photogenic function as a mating adaptation in the Photinini. *Can. Ent.* **43**, 399–406.

McDermott, F. A. (1914). The ecologic relations of the photogenic function among insects. *Z. wiss. Insekt Biol.* **10**, 303–307.

McDermott, F. A. and Crane, C. G. (1911). A comparative study of the structure of the photogenic organs of certain American Lampyridae. *Am. Nat.* **45**, 306–313.

McElroy, W. D. (1947). The energy source for bioluminescence in an isolated system. *Proc. Natn. Acad. Sci., U.S.A.* **33**, 342–345.

McElroy, W. D. (1951). Properties of the reaction utilizing adenosine-triphosphate for bioluminescence. *J. biol. Chem.* **191**, 547–557.

McElroy, W. D. (1964). Insect bioluminescence. In *The Physiology of Insecta,* Vol. 1. (M. Rockstein, ed.), pp. 463–508. Academic Press, New York.

McElroy, W. D. and Coulombre, J. (1952). The immobilization of adenosine triphosphate in the bioluminescent reaction. *J. cell. comp. Physiol.* **39**, 475–485.

McElroy, W. D., Coulombre, J. and Hayes, R. (1951). Properties of firefly pyrophosphatase. *Archs Biochem.* **32**, 207–215.

McElroy, W. D. and Hastings, J. W. (1955). Biochemistry of firefly luminescence. In *The Luminescence of Biological Systems* (F. H. Johnson, ed.), pp. 161–198. A.A.A.S., Washington, D.C.

McElroy, W. D. and Hastings, J. W. (1957). Initiation and control of firefly luminescence. In *Physiological Triggers and Discontinuous Rate Processes* (T. H. Bullock, ed.), pp. 80–84. American Physiological Society Publications, Washington, D.C.

McElroy, W. D., Hastings, J. W., Coulombre, J. and Sonnefeld, V. (1953). The mechanism of action of pyrophosphate in firefly luminescence. *Archs Biochem. Biophys.* **46**, 399–416.

McElroy, W. D. and Seliger, H. H. (1965). *Light: Physical and Biological Action,* 417 pp. Academic Press, New York.

McElroy, W. D. and Strehler, B. L. (1949). Factors influencing the response of the bioluminescent reaction to adenosine triphosphate. *Archs Biochem. Biophys.* **22**, 420–433.

Osten-Sacken, Baron (1861). Die amerikanischen Leuchtkafer. *Stettin. ent. Ztg* **22**, 54–55.

Palay, S. L. (1957). The fine structure of the neurohypophysis. In *Ultra-structure and Cellular Chemistry of Neural Tissue* (H. Waelsch, ed.), pp. 31–49, Cassel and Co., London.

Peterson, M. K. and Buck, J. (1968). Light organ fine structure in certain asiatic fireflies. *Biol. Bull. mar. biol. Lab., Woods Hole* **135**, 335–348.

Rall, T. W. and Sutherland, E. W. (1962). Adenyl cyclase. II. The enzymatically catalysed formation of adenosine-3′,5′-phosphate and inorganic pyrophosphate from adenosine triphosphate. *J. biol. Chem.* **237**, 1228–1232.

Seaman, W. H. (1891). On the luminous organs of insects. *Proc. Am. Soc. Microsc.* **13**, 133–162.

Seliger, H. H., Buck, J. B., Fastie, W. G. and McElroy, W. D. (1964). Flash patterns in Jamaican fireflies. *Biol. Bull. mar. biol. Lab., Woods Hole* **127**, 159–172.

Smalley, K. N. (1965). Adrenergic transmission in the light organ of the firefly, *Photinus pyralis. Comp. Biochem. Physiol.* **16**, 467–477.

Smith, D. S. (1963). The organization and innervation of the luminescent organ in a firefly, *Photuris pennsylvanica* (Coleoptera). *J. Cell. Biol.* **16**, 323–359.

Smith, D. S. (1965). Synapses in the insect nervous system. In *The Physiology of the Insect Central Nervous System* (J. E. Treherne and J. W. L. Beament, ed.), pp. 39–57, Academic Press, London.

Snell, P. A. (1932). The control of luminescence in the lampyrid firefly, *Photuris pennsylvanica*, with special reference of the effect of oxygen tension on flashing. *J. cell comp. Physiol.* **1**, 37–51.

Sutherland, E. W. and Cori, C. F. (1951). Effect of hyperglycemic-glycogenolytic factors and epinephrine on liver phosphorylase. *J. biol. Chem.* **188**, 531–543.

Townsend, A. B. (1904). The histology of the light organs of *Photinus marginellus. Am. Nat.* **38**, 127–515.

Verworn, Max (1892). Ein automatisches centrum für die lichtproduction bei *Luciola italica L. Zentbl. Physiol.* **6**, 69–74.

Williams, F. X. (1916). Photogenic organs and embryology of lampyrids. *J. Morph.* **28**, 145–207.

Wood, R. W. (1939). A firefly "spinthariscope". *Science, N.Y.* **90**, 233–234.

Postembryonic Development and Regeneration of the Insect Nervous System

JOHN S. EDWARDS

*Department of Zoology, University of Washington,
Seattle, Washington, U.S.A.*

I. INTRODUCTION

If development is behavior (Edwards, 1967a), and behavior is physiology (Kennedy, 1967), we may usefully dispense with such arbitrary distinctions and thus justify this paper on development in a series devoted to physiology.

It must be admitted at the outset that progress in building a theory of insect development in terms of hormonal sequences that regulate the behavior of the integument has outstripped understanding of mechanisms of internal organ development, perhaps because the latter seem to have more in common with those of other Metazoa, and thus to offer less of special interest. Developmental changes in the anatomy of the nervous and muscular system of many insects have been described, but apart from the experimental studies of muscle development and regression such as those of Nuesch (1968), Lockshin and Williams (1965) and of Sahota and Beckel (1967a,b) for example, which have given some insight into the processes of development and meta-morphosis in the musculature, rather little is known of the morphogenetic mechanisms in either muscle or nerve, and of these, the nervous system is the less understood.

It is about 200 years since Lyonet's (1762) detailed dissections of the larval and adult nervous systems of *Cossus* revealed the extent of the reorganization that accompanies metamorphosis and about 100 years since Weismann (1864) examined histological changes during metamorphosis. Since the mid 19th century over 100 papers have been devoted solely or primarily to the description of post-embryonic development of the nervous system in insects, appearing with peak frequencies in the 1880's, the 1920's and the present decade, to give the impression of a 40-year cycle. The recurrent questions that explicitly or implicitly underly this descriptive work are these:

1. Does the adult nervous system of Holometabola add to, or replace, the larval nervous system? If the latter, how is the larval system dismantled?

2. If the adult nervous system contains new elements, where do they come from?

3. If larval neurons are used in constructing the adult nervous system, are their synaptic fields changed?

This review will attempt to show how far answers can be given to the first two questions; the third remains unanswered. The interesting historical background of the subject cannot be presented here, although such a project seems not to have been done, and would be timely. Instead, an attempt is made to consider representative studies from the more recent literature, many of which give access to earlier work. Some important aspects of postembryonic development of the nervous system are not considered, for example, the development of the stomatogastric and retrocerebral systems, the differentiation of receptors, the biochemistry of the metamorphosing nervous system and hormonal aspects of nervous system development. Nuesch (1968) has recently discussed the important role of the insect nervous system in myogenesis and this topic will not be included here.

Quite apart from its intrinsic interest, the postembryonic development of the insect nervous system is especially relevant to recent developments in neuroembryology and questions of the nature of neural specificity. Advances in knowledge of the ontogeny of the nervous system, which have been made almost exclusively with Vertebrates, stress the capacity of differentiating neurons for forming complex and specific connections during early stages. Thus, the means by which cells recognize each other and are critical in neuroembryology (Jacobson, 1966; Hughes, 1968). Recent work with insects (Drescher, 1960; Guthrie, 1967; Jacklet and Cohen, 1967a,b; Edwards and Sahota, 1968) has shown their potential for such studies and it is hoped that this review will illustrate further aspects of postembryonic development which are relevant to the general question of neural specificity.

II. PATTERNS OF DEVELOPMENT IN THE NERVOUS SYSTEM

A. ANATOMICAL AND VOLUMETRIC CHANGES

Analysis of absolute and relative growth patterns by the Munster school (Rensch, 1959; Neder, 1959; Hinke, 1961; Lucht-Bertram, 1962; Korr, 1968) using a volumetric approach first applied to insects by Hanstrom (1926a), and many other studies (Power, 1952; Richard and Gaudin, 1959; Panov, 1957, 1959, 1960a, 1961a, 1966; Satija and Kaur, 1966a,b) show that there is a wide range of patterns of brain growth. In general, the growth of the brain of Hemimetabola bears a constant negative allometric relation to whole body growth during postembryonic development. Representative figures for Hemimetabola expressing brain volume as a power function of body weight are 0.47–0.58 for the brain of several Blattaria (Neder, 1959). In Holometabola, brain growth is allometrically negative with respect to the whole body through the larval stages, with major expansion in the pupal stage, as in *Drosophila,* or in late larval life as in *Culex* (Hinke, 1961), *Myrmeleon* and *Apis* (Lucht-Bertram, 1962) and others (e.g. *Galleria* (Sehnal, 1965), *Danaus* (Nordlander and Edwards, 1968b)). A wealth of data in these papers on the relative growth of various brain centers demonstrates a range of patterns, many of which correlate well with the development of sense organs. But since the relation between volume of neuropile or brain centers and their role in brain function is not well understood, and frequently problematic, as with the great corpora pedunculata of *Limulus* (Hanstrom, 1926b) and the contrasting corpora pedunculata of *Japyx* and *Petrobius* (Hanstrom, 1940) for example, volumetric studies alone give a limited insight system. Patterns of increase in neuron numbers (Neder, 1959; Hinke, 1961; see section V C below) and glia (Gymer and Edwards, 1967) provide more significant information, but this is arduous work that is not helped by the small size and close packing of the majority of cell bodies in the central nervous system and has seldom been attempted. A good foundation has been laid for the analysis of the honeybee brain in these terms by Lucht-Bertram's volumetric study, and Witthoft's (1967) detailed counts of neuron and glia populations in the adult bee brain.

In the terminal ganglion of the house cricket, in which no new neurons are added during postembryonic growth, the 40-fold increase in volume bears a typical negative allometric relation to body growth and is accounted for largely by a 17-fold increase in glial cell number and an increase in cell volume of all components. Cortical volume exceeds neuropile volume in growth rate during embryonic and early postembryonic stages, but as neurons differentiate, the relationship is reversed. In the terminal ganglion of the cricket, for example, the most rapid relative increase of neuropile material occurs during the first four instars (Gymer and Edwards, 1967) while in the *Drosophila* brain neuropile growth overtakes cortical development in the prepupal and pupal period (Power, 1952).

Whatever the relative growth rates of the central nervous system, and its parts, it seems that growth in volume is smooth and not locked into the molting cycle (Power, 1952; Panov, 1962, 1963) but this conclusion should be accepted with caution, for it is known that glial cells show cyclic patterns of mitosis in *Acheta* (Panov, 1961) and DNA synthesis in *Danaus* (Norlander and Edwards, 1969a), though neuroblast and preganglion cell division patterns are not related to the molt cycle (see Section III B). Cyclic incorporation of nucleotides also occurs in the ventral nerve cord of *Thermobia* (J. A. L. Watson, personal communication). It remains to be determined whether synthesis of new neuroplasm varies in relation to the molting cycle.

B. ANATOMICAL CHANGES IN THE CENTRAL NERVOUS SYSTEM

The developing central nervous system not only changes in volume; there are also changes in the number of ganglia, which are achieved by loss of connectives and fusion of ganglia. The brain and sub-oesophageal ganglion are composite ganglia formed early in embryonic development. The ventral series of ganglia are derived from the primary segmental series by fusion in embryonic or in postembryonic development.

Hemimetabolan ganglia undergo fusion which may be partial as in the Orthoptera (Roonwall, 1937, *Locusta*) or extreme as in the embryonic fusion of all abdominal ganglia in some Hemiptera (Springer, 1967). The temporal pattern of fusion varies among related groups. In the house cricket *Acheta (Gryllus) domesticus,* for example (Panov, 1966), the abdominal ganglia occupy their primary segmental positions until egg development is two-thirds complete. Then the first to sixth abdominal ganglia each move forward one segment, the first fusing with the metathoracic ganglion, the original second abdominal fusing also before the time of hatching. The seventh to tenth abdominal ganglia progressively fuse during embryonic development to form the compound terminal ganglion. In the mole cricket *Gryllotalpa,* migration is more extensive and proceeds during post-embryonic development. Panov points out difficulties in explanations of ganglion fusion based on a form of neurobiotaxis (e.g. Illies, 1962).

Details of the spectacular anatomical changes of Holometabola were described long ago (e.g. Newport, 1832, Lepidoptera; Weismann, 1864, Diptera; Brandt, 1879, Coleoptera). Ventral nerve cord shortening and ganglion fusion may proceed during larval life in some forms (Cody and Gray, 1938; Menees, 1961) but the most striking changes occur during metamorphosis and these have been examined in some detail in *Galleria* by Pipa (1963, 1967, Pipa and Woolever, 1964, 1965) who has sought the mechanisms involved, having found Murray and Tiegs' (1935) explanation of ganglionic fusion in *Calandra oryzae* in terms of cell proliferation and overgrowth inapplicable to *Galleria.* During the pupal stage, the mesothoracic and metathoracic ganglia of

Galleria become contiguous and the ganglia of abdominal segments 1 and 2 fuse with the metathoracic ganglion. During this process connectives shorten to give a final cord length 15–20% shorter than that of the larva.

The shortening of the connectives is accompanied in *Galleria* by a coiling and looping of the axons within the neurilemma, and on this evidence, Pipa considers that the shortening cannot be achieved by an active shortening of axons, but rather must be brought about by contraction of glial elements. The ultrastructural picture of the central nervous system during metamorphosis (Pipa and Woolever, 1964) suggests that glial cells may unwrap their elaborate foldings and withdraw from interaxonic spaces during the reformation of the nervous system.

Humoral regulation of this response was implicated in experiments in which nerve cords were transplanted into host larvae (Pipa, 1967). The shortening process proceeds most rapidly in these isolated cords during the normal shortening of the host cord. Clearly shortening of the nerve cord is not initiated by the morphogenetic changes of the epidermis since shortening can occur in these isolated cords, but Pipa suggests that the altered pattern of shortening in isolated cords may be related to the loss of peripheral contact and restraint.

In another Lepidopteran, *Pieris,* a similar 30% shortening of the cord occurs (Heywood, 1965) but quite different mechanisms are proposed to account for the contraction; no coiling or looping of axons was seen, and the contraction follows the shortening of the body by a few hours. It is suggested that resorption of axonic material leads to shortening, thereby "pulling" the cell bodies forward from their original position in abdominal ganglia to the fused thoracic complex within a sleeve of glial cells. The elastic neural lamella would shorten passively. It seems unlikely that so similar and basic a change in two Lepidoptera should be achieved by so very different mechanisms. While some axons are shortening others must be lengthening, for peripheral nerves retain connection with the ventral nerve cord via the vacated shells of the first and second abdominal ganglia (Heywood, 1965). As if this were not sufficient complexity, it should be remembered that these nerves are involved in the differential preservation of abdominal musculature (Finlayson, 1956, 1960) and that they continue to conduct impulses at the time when the investing perineurial cells are in morphogenetic turmoil and seemingly unfit to isolate the nerves from the hemocoel.

III. THE SOURCE OF NEURONS AND GLIA

A. EMBRYONIC

Early embryonic development of the insect nervous system, most recently examined by Malzacher (1968), lies outside the scope of this paper and the formation of the primary segmental ganglia will not be discussed, but some observations regarding the embryonic origin of neurons and glia are relevant. It may be noted that most studies of insect neuroembryology have been purely descriptive or have approached the question of head segmentation and homologies, for which the nervous system seems to be of dubious value (Bullock and Horridge, 1965; Ullmann, 1967).

Neuroblasts differentiate early in embryonic development (Johannsen and Butt, 1941; Malzacher, 1968) and are readily recognizable by their size and staining properties. Proliferation of neuroblasts is accomplished by divisions in which the progeny are of equal volume. The production of neurons begins with an asymmetrical neuroblast division, yielding a daughter neuroblast, and a smaller, preganglion cell with is distinguished as a ganglion mother cell (Schrader, 1938). In the embryo repeated divisions, always in a plane perpendicular to the surface of the developing nerve cord, yield parallel columns of preganglion cells which, at least initially, are remarkably regular in position; indeed Carlson (1961) was able to identify single cells according to row and column over a period of several days in embryos of *Chortophaga viridifasciata,* and could thus trace the progeny of individual neuroblasts in radiation studies (Carlson and Gaulden, 1964). Older cells lie progressively farther from the neuroblasts which retain their peripheral position.

The symmetrical division of ganglion mother cells to form equal sized daughter cells (preganglion cells) follows a pattern first described by Bauer (1904), and subsequently by Schrader (1938), Panov (1963) and others in Lepidoptera and is similar in other groups (e.g. *Drosophila,* Poulson, 1956; *Tenebrio,* Ullmann, 1967; *Carausius* and *Periplaneta,* Malzacher, 1968). Several early studies (e.g. Heymons, 1895) and more recent works (e.g. Roonwall, 1937; Shafiq, 1954) do not describe the division of ganglion mother cells, but there seems no reason to doubt that the process is general and uniform throughout the Insecta. The number of divisions undergone by embryonic preganglion cells is not known with certainty; it may be more than one in some cases, but is limited to one in *Carausius* and *Periplaneta* according to Malzacher (1968).

Glial elements arise from preganglion cells as in the Vertebrate. Perineurial cells (glial cell *i* of Wigglesworth, 1959) arise from the medial strand cells of the ventral nerve cord in *Drosophila* (Poulson, 1956) and other species, from lateral preganglion cells as in *Locusta* (Roonwall, 1937) or from both sources as in *Tenebrio* (Ullmann, 1967). They are, without doubt, of neural origin and do not invade the nervous system from elsewhere. The differentiation of glial cells and neurons from a common source is of interest in considering the subsequent behavior of the cells for, as will be discussed below, the pattern of proliferation of glial cells is in phase with the molting cycle while neuron production proceeds quite independently of the molting cycle. It may be that glial cells respond to the humoral cycle as do cells of the integument even though their cycles are not necessarily simultaneous, but the restriction of glial multiplication to the period around the molt perhaps has a functional significance related to varying metabolic requirements of the nervous tissue in the intermolt and molting periods.

Pycnotic nuclei and chromatic droplets occur among preganglion cells of *Tenebrio* embryos, and are most abundant about 30–50 hours after the initiation of preganglion cell production (Ullmann, 1967). The occurrence of cell death at this stage is seemingly at odds with the generation of orderly cell arrays, and it would be of interest to know whether any regularity can be found in the extent and distribution of cell death in the embryonic nervous tissue at this stage.

The majority of neuroblasts disappear toward the end of embryonic development, leaving only small groups in the larval brain. In *Antheraea pernyi*, for example, four pairs, associated with the corpora pedunculata and the olfactory center persist (Panov, 1960, 1963). The pattern is similar in other Lepidoptera, for example in *Ephestia* (Schrader, 1938) and *Danaus* (Nordlander and Edwards, 1969b). The nonpersistent embryonic neuroblasts are reported to die in some cases, as in *Vespa crabro* (Bauer, 1904), *Antheraea* (Panov, 1960, 1963) or, as described by Malzacher (1968), who found no degenerating neuroblasts in *Carausius* and *Periplaneta* embryos, they may themselves become ganglion cells.

B. POSTEMBRYONIC DEVELOPMENT

1. Neurons

In most Hemimetabola it seems that only the brain retains grouped or scattered neuroblasts during postembryonic development. Neuroblasts have degenerated in all segmental ganglia of *Acheta (Gryllus)* *domesticus* before hatching occurs. The neuroblasts of the abdominal ganglia degenerate first, while those of the thoracic segments and the head ganglia are still dividing. Those of the thorax subsequently degenerate, and finally those of the brain, except for two isolated neuroblasts in the deutocerebrum and groups in the corpora pedunculata and optic lobes which continue to divide throughout postembryonic development (Panov, 1957, 1960, 1966; Neder, 1959; Drescher, 1960). The process is similar, but the timing different, in *Gryllotalpa* where loss of segmental neuroblasts is delayed into early postembryonic development (Panov, 1966). The persistence in postembryonic life of groups of dividing neuroblasts in association centers of the brain after the major afferent and efferent pathways have been laid down may be comparable with the post-natal proliferation of microneurons in the brain of mammals (Altman and Das, 1965). Perhaps in the insect also the postembryonic recruitment of globuli cells in the corpora pedunculata, olfactory centers, and optic lobes allows for some modulation in synapse formation as the developing animal experiences the environment.

Counts of cell numbers during postembryonic growth of the terminal ganglion of *Acheta* (Gymer and Edwards, 1967) show that there is a constant population of about 2100 neurons throughout postembryonic development though it is not yet established that all are functional throughout development. Glial cells, on the other hand, increase in number from 3400 to 20,000 during postembryonic development. The full giant fiber population of the ventral nerve cord of *Acheta* is present at hatching (Edwards, 1967a) and the increase in volume during postembryonic development is unaffected by altering the peripheral load. Johannson (1957) observed no increase in number of neurons in the postembryonic growth of the thoracic ganglion of *Oncopeltus.* Thus, leaving higher brain centers aside, it seems that the architecture of the hemimetabolan central nervous system is laid down in embryonic life, and the neuron population remains constant throughout development. This is not to say that cell size remains constant: neuron cell bodies, nuclei, axons and glia do increase in volume (e.g. Sztern, 1914; Trager, 1937; Neder, 1959; Edwards, 1967b).

In the Holometabola very great changes occur in the nervous system during postembryonic development which involve the addition of large numbers of neurons. Although Sanchez (1925) saw few mitoses and neuroblast divisions during the metamorphosis of the

brain of *Pieris,* and Hanstrom (1925) considered the entire comple-
ment of adult neurons to be present in undifferentiated form in the
larva, there can be no doubt that the general pattern of postembryonic
development and metamorphosis involves extensive production of
new neurons, both in the brain, as shown by numerous studies (e.g.
Schrader, 1938; Nordlander and Edwards, 1968a, 1969a), and in
thoracic ganglia (Heywood, 1965). Interneurons and motorneurons
degenerate in the pupa, but according to Panov (1963) many neurons
in the thoracic ganglia persist from larva to adult.

The adult brain of Holometabola is formed from persistent postembryonic neuroblasts. Certain cells of the embryonic nerve cord fail
to differentiate to the definitive neuroblast stage in the egg but
persist either as aggregates, as in the optic lobe anlagen (imaginal
epithelia) or as scattered cells. Neuroblasts persist in the olfactory
center and corpora pedunculata as compact groups and as scattered
single cells (Schrader, 1938; Panov, 1960; Nordlander and Edwards,
1969a). Cells of the optic lobe anlagen enlarge and take on the
cytological appearance of neuroblasts during postembryonic development, and undergo both equal division, thus increasing the neuroblast
population, and asymmetric division to give larger daughter neuroblast cells and smaller ganglion mother cells which in turn divide
to form preganglion cells, as products of mitoses in a plane perpendicular to the brain surface, just as in the embryo. Preganglion
cells divide an unknown number of times before differentiating to
become neurons or glia. Their behavior is thus strictly comparable to
that of the embryonic neuroblasts, and the similarity extends to the
occurrence of cell death in both neuroblasts and ganglion cells during
development of the adult brain.

In general, the development of the optic lobes adds the greatest
number of new cells but these are not the only regions that undergo
change. The functional neurons of the larval corpora pedunculata, for
example, must work in an environment that is sometimes seething
with differentiating ganglion cells derived from neuroblasts associated
with these centers. The temporal patterns of change in some brain
centers are discussed in section V. The differentiation of motor and
interneurons in the central nervous system requires further attention:
Panov (1963) has rejected earlier distinctions between larval and adult
neurons on a size basis, and regards the larger cells as motor-neurons
and the smaller as interneurons, but a difficulty is raised by Chiarodo's
(1963) observation that the small cell population rather than large cells
contribute to the 27% reduction in cell number in the mesothoracic

region of the adult ventral nerve cord caused by extirpation of the mesothoracic leg bud in the last instar.

2. Glia

Glial cells of all four types distinguished by Wigglesworth (1959) increase in volume and number during postembryonic development of Hemimetabola and Holometabola. The glial cell population of the terminal ganglion of *Acheta domesticus* may be taken as an example. Counts of glial cells, excluding glia *i* of the perineurial sheath, made in each instar indicated that an initial population of about 100 in the first instar ganglion increases to 17,000 in the adult. The ratio of neurons to glia thus decreases during postembryonic development from 1:0.5 to 1:8 during a forty-fold increase in ganglion volume (Gymer and Edwards, 1967). The pattern of division was not examined in this study, but Sztern (1914) found cell division in the mesothoracic ganglion of *Sphodromantis* and Panov (1962) found the proliferation of glia in *Acheta* to follow a division pattern that differs from the epidermis but which is nonetheless correlated with the molt cycle, the peak of division occurring during the last quarter of the last instar. Nor were divisions found immediately following the molt to the last instar, or before the final molt.

A cyclic pattern of mitotic activity was found during postembryonic development in the monarch butterfly *Danaus plexippus* in which tritiated thymidine is incorporated by glial cells for about twelve hours before and after larval ecdyses, persistent incorporation occurring among glia for about two days after the pupal molt (Nordlander and Edwards, 1969a). The behavior of each of the various glial cell types is considered below.

Glial *ii* and *iv* cells increase in volume and number during postembryonic development; all larval cells apparently persist through metamorphosis, at least in Lepidoptera, for no pycnotic nuclei were seen among these cells in the pupal nervous system of *Antheraea* (Panov, 1960b), *Pieris* (Heywood, 1965) and *Danaus* (Nordlander and Edwards, 1969a).

Glial *ii* and *iv* cells of the growing optic lobes appear to originate from cells dispersed among differentiating ganglion cells and are at first indistinguishable from them: they evidently originate from the same source as neurons in *Danaus*.

Glial *iii* cells, the giant glial cells, increase in size but not in number during the larval stages of brain development in *Danaus*. DNA synthesis occurs regularly during the period before and after each molt, and additional nucleoli appear. Cell division occurs only during a short period beginning shortly before pupation and extending for two days afterward. The mitotic figures of these cells, first figured by Bauer (1904), are bizarre by comparison with neighboring cells and they are clearly polyploid. Their behavior resembles that of endomitotic fat body cells described by Risler (1954).

There can be little doubt that earlier reports (e.g. Bauer, 1904; Umbach, 1934) of the entry of exogenous cells to the central nervous system to provide new glia are in error. Schrader (1938), Panov (1963), Heywood (1965) and our studies with *Danaus* agree that glial cells divide extensively, particularly during metamorphosis, and their origin has been traced to neuroblasts in *Antheraea* (Panov, 1963), for example, and in *Danaus* (Nordlander and Edwards, 1969a). There are some minor variations with respect to the time of division, but these are probably of no great significance. A detailed study of the proliferation of these cells would be illuminating; the division of a cell whose morphology during its functional phase is highly elaborate, often with profuse laminate processes, raises questions as to the mechanics of mitosis which would repay ultrastructural observations. Can they divide without interrupting their trophic function?

IV. THE PERINEURIUM

A. ORIGIN

Before blastokinesis in the embryo of *Antheraea pernyi,* the cells of the central nervous system are not enclosed within a neural lamella (Panov, 1960b). Only after reaching this point in development do peripheral cells of the ganglia flatten and differentiate to become perineurial cells (glia *i* of Wigglesworth, 1959). There are few divisions among these cells in the embryo, but their postembryonic pattern of proliferation is very similar to that of glia *ii* and *iv.* Median strand cells of the embryo in other species may partially (e.g. *Tenebrio,* Ullmann, 1967), or completely (e.g. *Drosophila,* Poulson, 1956) provide the perineurial cells.

B. METAMORPHOSIS

The perineurial cells divide in *Danaus* during a brief period before and after the molt as in *Antheraea* (Panov, 1960b) but from the beginning of the fifth instar divisions occur continuously until the fourth day of the pupal period, declining from the second to the fourth. The perineurium thickens during the third instar, especially around the developing optic areas. The cells change shape from squamous to columnar or cuboidal, and finally become multilayered. By the end of the last instar the perineurium has become thicker and spongy in texture. Among its cells many trachea and vacuolar cells, probably hemocytes, appear, then during the last day of larval life perineurial cells begin to disintegrate. Pycnotic nuclei occur in the perineurium of *Pieris* before pupation and a postero-anterior, dorsoventral gradient in the progress of perineurial cell death is seen in the early pupa (Heywood, 1965). During the early pupal stage it disintegrates and all but three or four layers of cells are lost. The residual perineurial cells secrete a new neural lamella, and finally form a single cell layer below it.

The neural lamella also begins to disintegrate during the last day of the last larval instar (Panov, 1961b, 1963; Ashhurst and Richards, 1964; Pipa and Woolever, 1964, 1965; Nordlander and Edwards, 1968b, 1969a). The process of demolition of the neural lamella has been described many times, especially among Lepidoptera, and there are differences of interpretation, especially of the role of hemocytes and phagocytosis. In *Galleria* (Ashhurst and Richards, 1964; Pipa and Woolever, 1964, 1965) the neural lamella separates from the perineurial cells about twenty-five hours before hemocytes invade the cavity so formed. These cells appear to be involved in the breakdown of the neural lamella about the brain and the ventral cord and have also been observed in *Ephestia* (Schrader, 1938), *Danaus* (Nordlander and Edwards, 1969a) and *Pieris* (Sanchez, 1924, although Heywood (1965) described fragmentation and autolysis of the perineurium in the ventral cord of *Pieris* without the intervention of hemocytes. Panov (1963) considers the hemocytes that penetrate the nerve cord of *Antheraea* at the time of neural lamella breakdown to act as phagocytes, ingesting fragments of neural lamella and dead cells, though the majority of cells disintegrate without phagocytosis. Pipa and Woolever (1965) report the presence of ingested fibrillar material in electron micrographs of cells, presumably hemocytes, within the neural lamella, and conclude that these cells ingest fragments of degenerating lamellar material by phagocytosis.

The neural lamella is lost during early metamorphosis in Lepidoptera and possibly all Holometabola in which there is extensive rearrangement of the nervous system, and during this period the cord is sheathed by remnant perineurial cells and glia. The nerve cord is extremely fragile, and the metamorphosing pupa shows the lowest level of responsiveness at this time. It seems also to coincide with the period of minimal electrical activity in the central nervous system, at least in non-diapausing species such as *Galleria*. The neural lamella is reconstructed, probably entirely by remaining perineurial cells after the major anatomical changes associated with metamorphosis are complete.

V. POSTEMBRYONIC DEVELOPMENT OF THE BRAIN AND SENSE ORGANS OF THE HEAD

Description of the postembryonic development of brain centers in many insects began with major contributions in the latter part of the last century and early this century, e.g. Weismann (1864), Viallanes (1882 and later papers), Henneguy (1903) and Bauer (1904). Details and variations were added by subsequent description and some experimental studies, for example that of Kopec (1922), opened up questions of the morphogenetic relationships between brain and sensory structures in development. More recently an important series of papers by Panov (1957, 1959, 1960a, 1961a,b, 1966) and studies by Neder (1959), Afify (1960), Hinke (1961) and Lucht-Bertram (1962) have brought some pattern into the many observations on brain development, and against this background some progress has been made in analyzing cell behavior during growth of brain centers (Nordlander and Edwards, 1968a, 1969a,b,c).

The object of this section is to outline the development of the better known centers in the light of these studies, and to consider experimental work on the morphogenetic relationship of the eye and antennae with their respective brain centers, the optic lobe and the olfactory center.

A. EYE AND OPTIC LOBE

The receptors and afferent fibers of both ocelli and compound eyes are derived from epidermal cells which may differentiate in the embryo or during post-embryonic development. Mechanisms of receptor differentiation lie outside the scope of this review but it is worth noting that the differentiation of the lens and associated cells of the dioptric system have provided experimental material for advances in concepts of gene action (review: Bodenstein, 1953; Kaji, 1960 and earlier papers), and continue to receive attention as differentiating systems (Imberski, 1967; White and Sundeen, 1967) well suited to cytological studies.

1. The Compound Eye of Hemimetabola

The compound eye of Hemimetabola functions from hatching and grows directly toward the adult state by increase in cell size and by adding few or many facets, depending on the species, from cells which proliferate at the border of the eye, without relation to the molting cycle (review: Bodenstein, 1953). These border cells are evidently capable of some degree of regulation, for compound eyes which regenerate after cautery of all ommatidia in early instars (Edwards, 1967b; Heller and Edwards, 1968) may approximate the normal eye in the adult, and have comparable receptive fields (Edwards *et al.*, unpublished).

Excised eyes of *Periplaneta* removed by cutting through the lamina ganglionaris survive and molt after transplantation to the ventral surface of coxae (Wolbarsht *et al.*, 1966), and second instar *Locusta* eyes excised between retina and lamina continue growth after rotation through 180° (Horridge, 1968).

2. The Compound Eye of Holometabola

The characteristic growth of the holometabolan eye involves explosive development during metamorphosis of the eye imaginal disk in which differentiation of cells destined to form ommatidia may begin during late larval life. Where the larva has a compound eye, as in the mosquitos *Culex* and *Aedes,* development procedes continuously (Haas, 1956) although mitotic activity in the developing eye placode is interrupted briefly at each ecdysis (White, 1961). In several species, representing early and late types of eye development, a similar pattern of differentiation, first described in detail by Umbach (1934), has been observed (Wolsky, 1956; White, 1961). Differentiation begins at a point on the prospective eye, usually postero-dorsal in position, which is critical in eye morphogenesis, and spreads as a wave of mitotic activity across the surface in an antero-ventral direction. This wave is followed by a wave of differentiation (White, 1961).

The behavior of the retinal neurons has received much less attention than that of the lens system, and the process by which connections are made with the optic lobe is in need of detailed examination. In Lepidoptera the larval visual system, the stemmata, send a slender nerve to the protocerebrum. At metamorphosis, the stemmatal nerve, which has become thickened during later larval life by sheath cells, probably glia, carries the remnants of stemmatal pigment toward the brain, where it remains permanently enclosed by the perineurial sheath. Products of the remaining sheath of cells from the old stemmatal nerve provide the framework on which the differentiating cells of the ommatidia grow centripetally. The stemmatal nerve sheath, in which many cells divide, splits at its hypodermal extremity to form branches, first in a dorso-ventral plane, then in the horizontal plane to form a radiating array of slender filaments connecting the hypodermis with the developing protocerebrum. The pattern described above for *Danaus* (Nordlander and Edwards, 1968a, 1969b) supports the early observations of Johansen (1892) rather than those of Bauer (1904) and Umbach (1934) who regarded the strands connecting the hypodermis and the optic lobes as inwardly migrating epidermal cells.

3. The Optic Lobe

Though there are many minor variations, the general pattern of optic lobe structure allows comparison of the major neuropile masses and their formation in a wide range of insects. The three successively more medial neuropiles (neuropiles I, II, III of Burtt and Catton, 1966) will be referred to here as the lamina, medulla and lobula following Bullock and Horridge (1965).

Adult optic lobe anlagen in Holometabola invaginate early in embryogenesis from the first and part of the second protocerebral lobes to form separate densely packed rods or plaques of minute cells in which the characteristic neuroblast-preganglion cell relationship is not at first evident.

Panov's (1960a) comparative work on the development of the optic lobes has brought order to apparently conflicting observations of earlier work. In the hemimetabolan optic lobe, a reduced post-embryonic population of neuroblasts continues to divide throughout postembryonic development. Both asymmetric and symmetric neuroblast divisions occur, only the former maintaining a constant orientation with respect to the periphery of the center. The addition of new ganglion cells is doubtless associated with the continued accession of new afferent fibers from the compound eye. Panov's observations on the optic lobe of *Tettigonia* (1960) follow a pattern described earlier in other Hemimetabola by Bauer (1904) in *Aeschna,* and by Graichen (1936) in *Nepa.*

The developmental pattern in Holometabola is determined by the complexity of visual structure in the larva, and varies between two extremes typified by the mosquito *(Culex),* the sawfly *(Pristiphora),* and campodeiform coleopteran larvae on one hand, all of **which have** active larvae with vision, and the honeybee *(Apis)* and many Lepidoptera on the other, in which the larval visual system is rudimentary.

In *Pristiphora pallipes,* the larval optic lobe has three neuropiles. Anatomical changes in the larval centers become visible in the late larval instars as new material appears in each of the neuropiles. In the prepupal stage a disintegration of neuropile and associated ganglion cells begins and is complete in the young pupa. The larval and adult neuropiles thus arise from neighboring but different neuroblasts (Panov, 1960a). In the bee, as in many Lepidoptera where the larval optic lobe is minimally developed, the adult optic lobe develops independently and continuously through postembryonic stages. Panov's (1960a) observations on the bee demonstrate this clearly and add to the pattern established by Schrader (1938) in *Ephestia.*

The outline of optic lobe development in *Danaus* (Nordlander and Edwards, 1969b) which follows, indicate the general features of the system for Holometabola in which the larval optic centers are rudimentary. It will serve as a basis for a consideration of details of cellular activity in the developing optic lobe.

Two layers are present in the optic lobe anlage of *Danaus*–the inner and outer optic lobe anlagen which form rods of cells about the larval optic centers. Development proceeds throughout postembryonic life, accelerating greatly at the time of pupation. The proliferation centers, and subsequently their cell progeny, occupy most of the head capsule lateral to the brain hemispheres. Early growth and expansion of the optic lobe anlagen accrues at first entirely from formation of new neuroblasts, but by the end of the second instar, the differentiation of optic ganglion cells and optic fiber masses contribute to the volume. By the early fifth instar, the outer anlage has lengthened to form a 1½ turn coil and at this time symmetric neuroblast divisions cease. Cells of the inner anlage undergo few equal neuroblast divisions, and its form and size change little during larval life. At the time of pupation, the outer anlage expands to form an open horseshoe, and from this point on, through the first third of pupal life, neuroblasts degenerate.

Asymmetric divisions begin during the second instar, the first ganglion mother cell divisions following within a day of the first unequal neuroblast divisions. The first appearance of fibers follows the division of ganglion mother cells by about 20 hours in *Danaus,* a relationship which parallels that of neuropile formation in the embryo of *Tenebrio* (Ullmann, 1967).

The three major neuropile regions of the optic lobe, the medulla, the lobula and the lamina are formed from the neuroblasts of the optic lobe anlagen. The medulla appears first in development as a ribbon-like fiber mass discernible in the second instar. As fiber proliferation proceeds, the ribbon widens in a direction perpendicular to the long axis of the anlage. Later in development, fibers of centripetal cells reach the medulla from the lamina and contribute to the increase in volume.

The lobula appears during the fourth instar as a small fiber region between the arms of the growing medulla. By the last instar, the anterior and posterior medulla are distinct and send fibers to the posterior lobula.

The lamina is the last fiber mass of the optic lobe to form. It arises as a result of the differentiation of a cap of cells formed by the lateral rim of the outer optic lobe anlage, by centripetal growth of fibers from the ommatidia and by the entry of a small number of fibers from the differentiating medulla. Fiber formation by cells in the lamina first occurs toward the end of the last larval instar and proceeds through the pupal period. During the third and fourth day after pupation the maximum accession of fibers from the ommatidia occurs.

The inner chiasma is formed during the fifth instar when fibers of the posterior lobula traverse the space between it and the anterior lobula. In crossing the space they intersect with fibers from the medulla which pass into and between the two lobulae. The external chiasma forms as the growing fibers of the lamina cortical cells cross one another before entering the medulla. They cross at $90°$ but later, as the fiber masses broaden, the angle increases to about $120°$.

General features of the sequence outlined for *Danaus* are shared with a variety of Lepidoptera though timing of the sequence varies widely; development of the optic lobes may be initiated in the pupae of *Pieris* (Hanstrom, 1925) and *Ephestia* (Umbach, 1934; Schrader, 1938) while in *Danaus* (Nordlander and Edwards, 1969b) as in the bee (Panov, 1960a) and *Drosophila* (El Shatoury, 1956) development proceeds throughout the larval period.

4. *Cell Movement and Differentiation in the Optic Lobe*

Radioautographic studies using timed injections of tritiated thymidine to label cells in proliferating parts of the brain, as has been done with vertebrate material by Levi-Montalcini (1963),

Sidman (1963) and others, have revealed something of the behavior and the source of cells involved in the morphogenesis of the several parts of optic lobe described above. Reconstruction of the movements of cells from the imaginal epithelium was made possible by comparing several series of radioautographs prepared at successive larval and pupal stages in each of which a standard series of intervals between injection and fixation was used (Nordlander and Edwards, 1968a, 1969b).

The inner optic lobe anlage contributes only the cortex of the posterior lobula which arise from strands of cells on the lateral ridge of the inner optic anlage. The formation of the medulla and lamina cortices from the outer optic lobe anlage is summarized in Fig. 1.

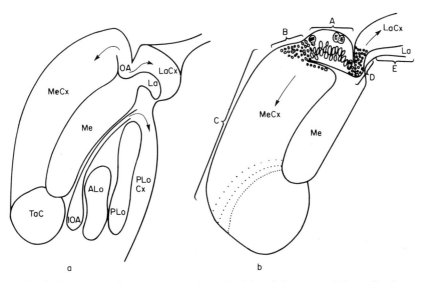

Fig. 1. Morphogenetic processes in the optic lobe of the monarch butterfly *Danaus plexippus.*

a. Diagrammatic horizontal section of early pupal optic lobe showing the sources of some components of the adult optic lobe, notably the lamina cortex and medulla cortex from the outer optic anlage, and the lobula cortex from the internal optic anlage.

b. Diagrammatic horizontal section showing in more detail the products of the outer optic anlage. In region A neuroblasts divide. Regions B and D include dividing ganglion mother cells. In regions C and E ganglion cells differentiate. Note that the lamina cortex and medulla cortex are formed from the lateral and medial borders respectively of the anlage.

Abbreviations: ALo, anterior lobula; IOA, internal optic lobe anlage; La, lamina; LaCx, lamina cortex; Me, medulla; MeCx, medulla cortex; PLo, posterior lobula; PLoCx, posterior lobula cortex; TaC, tangential cells. (From Nordlander and Edwards, 1969b.)

The cell bodies of the optic lobe begin to differentiate soon after they are formed and before there is any major change in the appearance of the cell at the light microscope level, other than enlargement of the nucleus. This is a neglected topic and requires further examination at the electron microscope level. Beier (1928) noted a progressive decrease in Nissl-like substance in the cytoplasm of differentiating cells of the optic cortex, which he compared to changes in the degenerating vertebrate axon. A similar change is seen in *Danaus*, where it is attributed to a dispersion of RNA as the cytoplasm increases in volume.

5. The Morphogenetic Relationships of Eye and Optic Lobe

The morphogenetic relationship between the compound eye and the optic lobe is only partially understood, despite the attention it has received in a number of studies reviewed by Bodenstein (1953) and by Pflugfelder (1958), and in recent work by Schoeller (1964).

It has been generally stated that the development of dioptric components of the compound eye is unaffected by their separation from the optic lobe. In *Pentatoma prasina* the compound eye grows normally without the optic lobe, and the embryonic development of eye and optic lobe are independent although they are closely correlated in postembryonic development in *Carausius* (Pflugfelder, 1958). Schoeller's (1964) work with *Sarcophaga* shows that the cornea, pseudocone and rhabdome develop normally after connections with the optic lobe have been permanently severed in the third instar, two days before pupation. In these preparations centripetal fibers from the retina reach beyond the fenestra, and terminate there without apparent orientation.

The eye of *Acheta* will regenerate after cautery in an early instar which can cause extensive damage to the optic lobe of the same side (Edwards, 1967b). The possibility that an intact optic lobe is necessary for normal eye development in *Bombyx* is suggested by Wolsky (1938); Drescher (1960) observed regression of the compound eye in *Periplaneta* after extirpation of one side of the brain. Removal of large segments of brain tissue may perhaps have consequences for neurosecretory activity that are not manifested in lesser operations, but the eye of *Periplaneta* can undoubtedly develop at a distance from the intact brain, for eyes severed through the lamina ganglionaris immediately below the fenestrated basement membrane and transplanted to the ventral coxal surface of the third leg of the seventh or eighth instar nymphs, survived and molted at least twice (Wagner and Wolbarsht, 1963; Wolbarsht *et al.,* 1966). Post retinal fibers from the transplanted eyes entered the leg nerve.

The volume of neuropile in the three successively deeper layers of the optic lobe is reduced in direct relation to the reduction of facet number in the eye of mutant *Drosophila* (Power, 1943; Hinke, 1961). In eyeless mutants studied by Power, hypoplasia was 100% in the lamina, 85% in the medulla, 59% in the lobula, and 57% in the lobular plate. In animals in which the centripetal fibers are present but fail to contact the optic lobe, the volume of the adult neuropile regions is strictly comparable to that of the eyeless flies. Eyeless mutants also show hypoplasia of the antennal centers (Power, 1946; Hinke, 1961). Other centers were apparently unaffected in Power's material, but Hinke found a small decrease in relative corpora pedunculata volume. Power noted a decrease in the number of monopolar cell bodies associated with the lamina in mutants with reduced ommatidial number, but it is not known if the reduction involved excessive cell death following a normal development of the optic lobe anlagen. A raised level of cell death during optic lobe development in such mutants would provide evidence for the dependence of developing interneurons of the optic lobe on the establishment of specific cell contacts in the developing neuropile. Power concluded that the observed hypoplasia is in direct relationship to the loss of centripetal fibers, but in *Sarcophaga* (Schoeller, 1964) none of the optic lobe neuropiles developed when connection with the eye was severed before metamorphosis. It is difficult to assess how much of the apparent variety in morphogenetic relationship between the eye and the optic lobe stems from experimental and observational differences. Despite the many studies mentioned above, and others (reviews: Bodenstein, 1953; Pflugfelder, 1958), the question remains open.

B. ANTENNAE AND THE OLFACTORY CENTER

As with the eye and optic lobe, the pattern of development of the antennae and the olfactory center is direct and progressive in the Hemimetabola, but varied in Holometabola, depending on the difference between larval and adult antennae: where larval antennae and olfactory centers are well developed, they are replaced during metamorphosis (Panov, 1961a).

1. Antennal Development in Holometabola

The antennae of the adult develop from the antennal disk during metamorphosis. The possibility that the differentiation of the antenna is under nervous control has been argued by Schoeller (1964) on the basis of implantation experiments with *Calliphora* in which antennal disks of third instar larvae were transferred to the abdomen of hosts of the same age. Nerve connections with the terminal abdominal nerve, or a branch of it, were invariably formed, and the degree of differentiation in the antenna proved to be correlated with the quantity of nerve in the implant. Schoeller regards these axons as centrifugal fibers, and concludes that differentiative processes in the antennal anlage are released only after early innervation by motor fibers. Proof of the centrifugal origin of these fibers and elimination of the possibility that they are centripetal fibers destined, in the course of normal development, to grow inward to the olfactory center of the brain, would provide strong evidence for the neural control of differentiation. If the case for neural control of differentiation of the antenna is proven, it would stand in contrast to eye development which Schoeller found to be independent of central control in *Calliphora*.

2. Olfactory Center Morphogenesis

In the Hemimetabola the early postembryonic olfactory center is identical in structure with that of the adult, except for volume (Panov, 1961a), although in *Locusta* (Afify, 1960) the characteristic glomerular structure is said to arise in the second instar. The glomeruli increase in volume but not in number during postembryonic development.

In holometabolous larvae with rudimentary antennae, the olfactory center is correspondingly suppressed, as in *Apis* (Lucht-Bertram, 1962; Panov, 1961a), *Myrmeleon* (Lucht-Bertram, 1962) and *Culex* (Hinke, 1961; Panov, 1961a). Extreme reduction occurs in some cyclorrhaphous Diptera: Panov (1961a) found no deutocerebral structure in newly hatched larvae of *Musca vomitoria*, and *Calliphora* similarly lacks a distinct olfactory center in early stages (Gieryng, 1965).

Panov distinguishes two developmental patterns in the olfactory center. In the honey bee, *Apis*, for example, the olfactory center develops from several isolated neuroblasts and a compact group, which are active up to the second pupal day. The neuropile is at first homogeneous, glomeruli not appearing until during metamorphosis when afferent fibers arrive from the developing adult antennae. The glomeruli of the adult olfactory lobe are thus composed of new neuropile derived in part from the proliferation of fibers during the early pupal period, and from afferent fibers of the adult antennae. *Calliphora* similarly lacks a glomerular antennal center until early in adult development (Gieryng, 1965).

Where larval olfactory centers are present, as in *Danaus* and *Tenebrio,* the larval and adult centers are separate structures. Three isolated neuroblasts form the proliferation center for the antennal center in *Danaus.* Neuroblast divisions begin at the end of the second instar. The larval center has a glomerular structure, but is lost at the end of the last instar, to be regained as the adult antennae develop on the third pupal day (Nordlander and Edwards, 1969c). The newly hatched *Tenebrio* larva has a glomerular olfactory center which is destroyed early in pupal life and is replaced by adult glomeruli (Panov, 1961a). Degeneration of cells associated with the larval center occurs during early metamorphosis. Degeneration studies have demonstrated the loss of volume in the olfactory center after amputation of antennae, though glomerular structure is retained (Titschack, 1928; Panov, 1961a). Removal of the adult antenna in the early pupal stage prevents the formation of glomeruli. Thus it appears that the afferent fibers are necessary for the formation of glomeruli, but not to maintain them.

C. CORPORA PEDUNCULATA

The development of the corpora pedunculata appears to be continuous in all insects, and proceeds in Holometabola without replacement and destruction of larval neurons, as occurs in the optic and olfactory centers. The range in complexity of corpora pedunculata structure, principally the number of calyces and lobes, cannot be considered in detail here; it is sufficient to observe that the center increases in complexity during postembryonic development in all species examined. Panov (1957, 1966) distinguishes two patterns of differentiation based on the number and distribution of neuroblast rudiments, and on the differentiation of neurons within cell groups; neuroblasts may occur as a compact and contiguous group in each calyx, as in Orthoptera, Hymenoptera and Coleoptera, while in other orders (e.g. Diptera and Lepidoptera) varying numbers of scattered neuroblasts form a proliferation center. In general, the adult corpora oedunculata of the first type are the more complex.

The corpora pedunculata of Hemimetabola develop in the embryonic brain, though the stage reached at hatching varies; in *Acheta* the primary calyx and peduncle are already present, but in *Gryllotalpa* the calyx does not appear until early in postembryonic life. In *Locusta,* similarly, the formation of calyces occurs in early instars (Afify, 1960). Cloarec and Gouranton (1965) describe further details of postembryonic differentiation of the corpora pedunculata in *Locusta.*

The range in total adult globuli cell population in the corpora pedunculata of three species of roach is remarkable; figures for adult males are as follows: *Periplaneta americana* 403,600; *Blatta orientalis* 314,100; and *Phyllodromia germanica* 73,360 (Neder, 1959). The globuli cell population of the corpora pedunculata in *Culex pipiens* adult males is 2110 (Hinke, 1961). The postembryonic increase in cell numbers in the four species examined is 5, 4, 7 and 20-fold respectively.

The greatest increase in volume in relation to brain volume occurs during the early instars in the cockroaches, and in the pupa and early adult development of the mosquito. There is a continual increase in the proportion of neuropile to cortical (globuli cell) volume from about 50% at hatching to about 120% in the adult, with the greatest relative increase in the second and third instars. The rate of increase in cell number is also greatest in the early instars.

In Holometabola the relative rate of corpora pedunculata development throughout larval stages varies considerably—in *Apis* (Panov, 1957) and *Danaus* (Nordlander and Edwards, 1969b), for example, development is continuous and more or less evenly distributed through postembryonic development, accelerating in late larval and pupal stages. Their growth in volume in *Apis* begins in the second instar and proceeds at different rates in drone, queen and worker to give adult populations of about 300,000 neurons in the drone, and 340,000 neurons in the worker (Witthoft, 1967). In *Myrmeleon* (Lucht-Bertram, 1962) and *Culex* (Hinke, 1961) corpora pedunculata begin development in the embryonic brain, and in these active larvae probably function from hatching on. In other insects, for example *Ephestia* (Schrader, 1938), growth of these centers progresses little before metamorphosis. The significance of the corpora pedunculata in larval behavior is unknown, but it is presumably of little importance in *Ephestia* since the neuroblast divisions leading to corpora pedunculata development do not begin until the last larval instar.

In *Danaus*, the corpora pedunculata are represented at hatching by two clumps of three neuroblasts in each brain half (Nordlander and Edwards, 1969c). The neuroblast population doubles during the first instar and increases more slowly thereafter to give about thirty cells per neuroblast center by the middle of the last instar. As many as 30% of the neuroblasts may be involved in symmetrical and asymmetrical divisions at any one time during early instars. Their pattern of division is apparently unrelated to the mounting cycle. The basic double stalked neuropile structure is established during the first instar, and calyces are first evident in the third instar. Asymmetric divisions yield ganglion mother cells which divide further, the progeny then differentiating to form globuli cells. Thymidine labeling shows that cells are displaced centrifugally as neuroblasts and their progeny proliferate in the cap of cells overlying the calyx. All neuroblasts degenerate during adult development.

D. THE CENTRAL BODY

The development of the central body, an association center, is closely correlated with that of photoreceptors. Both the dorsal glomerular part and the ventral part composed of transverse fibers are present at hatching in many Hemimetabola (Panov, 1959) and this form is retained throughout postembryonic development, both the glomerular and transverse parts adding new fibers. Details of central body development in *Acheta domesticus* and *Gryllotalpa* are discussed by Panov (1966).

The first appearance of the central body in Holometabola ranges from embryonic stages, as in *Tenebrio molitor, Antheraea pernyi* and *Culex pipiens* (Panov, 1959; Hinke, 1961) through larval stages, as in *Danaus plexippus* (Nordlander and Edwards, 1969c) and *Apis* (Lucht-Bertram, 1962) to the pupa as in *Ephestia kuhniella* (Schrader, 1938), *Calliphora* (Gieryng, 1965) and *Oryctes nasicornis* (Jawlowski, 1936). The correlation between development of the central body and presence of stemmata in larvae, first noted by Bretschneider (1914) is borne out by Panov's (1959) observations.

The central body is not identifiable in first and second instar of *Danaus*. In the third instar, swirls of fibers lie in the region to be occupied by the central body, which in the fourth instar becomes evident as it becomes delimited by glial *iv* cells, and in the larval instar it acquires a large number of fibers. Although Panov describes neuroblasts and subsequently cell growth associated with the central body, the comparable cells in *Danaus* and *Panorpa* (Bierbrodt, 1942) appear to be associated with the protocerebral bridge rather than the central body.

E. THE PROTOCEREBRAL BRIDGE

This region of the brain has been neglected from the developmental point of view although larval and adult structure has been compared in several cases (e.g. Hanstrom, 1925; Schrader, 1938; Bierbrodt, 1942). Differences in interpretation of neuroblast and cell groups in relation to the protocerebral bridge and the central body (e.g. Bierbrodt, 1942; Panov, 1959) emphasize the need for detailed neuroanatomical studies on this region of the brain. The protocerebral bridge is well developed in the newly hatched bee larva (Lucht-Bertram, 1962) where it forms 4.8% of the total brain, and in the newly hatched *Myrmeleon* larva. In *Danaus* the protocerebral bridge receives fibers from neighboring cells in the first instar. Scattered neuroblasts contribute further to the protocerebral bridge during larval development, and growth continues through the larval period. Connections with the optic lobe are evident from the outset; connections develop with the dorsal protocerebral neuropile in the second instar, and with the central body in the later larval instars. Transverse connections are not made until during adult development. In *Calliphora* the protocerebral bridge develops in the second half of the pupal stage (Gieryng, 1965).

VI. CELL DEATH IN THE DEVELOPING NERVOUS SYSTEM

Cell death plays a role in the development of the nervous system from the embryo (Ullmann, 1967) to the aging adult (Rockstein, 1950). The demise of neuroblasts follows such a regular pattern that it appears to be "programmed" and it would be of considerable interest to determine whether the life and productivity of a neuroblast may be altered by manipulating the humoral environment. Certainly they are capable of some regulation, for extra neurons are produced when cortical cells of corpora pedunculata of immature *Periplaneta americana* are removed (Drescher, 1960). The adult corpora pedunculata, in which the neuroblasts have degenerated, do not regenerate. Degenerating neuroblast nuclei form distinctive chromatic droplets as described in dying epidermal cells (Wigglesworth, 1942) and the nuclear envelope disappears. The proximity of dying and actively dividing neuroblasts is a striking feature of proliferating regions of the corpora pedunculata, as it is of the developing optic lobes (Nordlander and Edwards, 1968a).

Cell death is also the fate of some larval neurons. Observations of degenerating larval neurons have recurred frequently since Weismann (1864) reported the degeneration of larval elements. Larval optic and olfactory neuropiles are dismantled and are replaced by adult centers. Motor neurons and globuli cells degenerate in thoracic ganglia early in the pupal stage in Lepidoptera (Panov, 1963; Heywood, 1965). Questions remain as to which larval neurons persist into the adult, and as to whether they form new connections.

The mechanism of neuron destruction has been contested: Henneguy (1903) and Bauer (1904) both concluded that phagocytes are involved, though not in all cases. Cajal and Sanchez (1921) found historical evidence for phagocytic activity in *Bombyx* and *Pieris,* but Schrader (1958) and Panov (1963) described the destruction of larval neurons in *Ephestia* and *Antheraea* respectively without the intervention of phagocytes, and none appear during the degeneration of neurons in the antennal centers of *Danaus* during metamorphosis (Nordlander and Edwards, 1968). Hemocytes undoubtedly invade the central nervous system during metamorphosis, and the ultrastructural observations of Pipa and Woolever (1965) indicate that they are involved in eroding the neural lamella, but the balance of evidence supports the view that larval neurons are demolished at their own hand. Perhaps the signal to initiate this process is the termination of input from the obsolete larval receptors, and the mechanism may thus be comparable to the destruction of abdominal musculature in the adult saturniid moth (Lockshin and Williams, 1965).

Cell death also occurs among ganglion cells during neuron differentiation as in the optic lobe of *Danaus* (Fig. 2) (Nordlander and Edwards, 1968a, 1969b). This feature of neural development has not hitherto received attention. Pycnotic nuclei occur among optic ganglion cells during the height of differentiation. The peak frequency of preganglion cell division precedes that of pycnoses in ganglion cells by about three days, and it is evident from radioautographic studies that cell death is rare among ganglion cells at earliest stages of differentiation; at least some of the dying cells have axon hillocks and already send processes to the neuropile. Just how a particular cell acquires the stigma of redundancy is an intriguing question.

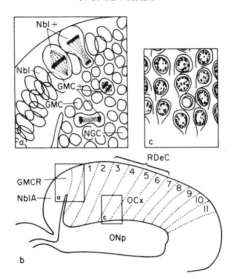

Fig. 2. The developing optic lobe of a mid-pupal brain as seen in horizontal section.
a. The region of neuroblast and ganglion mother cell divisions.
b. Diagram showing the developmental pattern of the optic lobe. Numerals indicate the approximate age, in days, of cells contained within each of the delineated areas of the cortex.
c. A degenerating cell among differentiating optic ganglion cells.
Abbreviations: GMC, ganglion mother cell; GMCR, ganglion mother cell region; GMC÷, ganglion mother cell division; Nbl, neuroblast; NblA, neuroblast aggregate; Nbl÷, neuroblast division; NGC, new ganglion cell; OCx, optic cortex; ONp, optic neuropile; RDeC, region of degenerating cells. (From Nordlander and Edwards, 1968a.)

There may perhaps be a parallel in the insect nervous system with the observations of Hamburger and Levi-Montalcini (1949) on spinal ganglia of chick embryos, and Hughes (1959, 1961) on the ventral horn cells of the *Xenopus* embryo, that degeneration of neurons may be related to the peripheral load encountered by outgrowing fibers and in the degeneration of cells in the superior colliculus of the mouse that occurs 4 to 7 days after eye removal at birth (de Long and Sidman, 1962). The failure to make appropriate connections with other cells, neurons or muscle cells, seems to decide the fate of these differentiating vertebrate neurons. It may be that the persistence of neurons in the highly structured optic lobe of the insect depends on the establishment of cell contacts, and that these determine the subsequent differentiation into one of the numerous neuron morphologies recognized by Cajal and Sanchez (1921). The degree of specificity of vertebrate neurons in the development and regeneration of the visual system (reviews: Sperry, 1965; Jacobson, 1966; Szekely, 1966) would certainly account for cell behavior of this complexity. The optic lobe of the insect may prove to be ideal material for such analyses.

VII. REGENERATION IN THE NERVOUS SYSTEM

A. THE BEHAVIOR OF REGENERATING NEURONS

The pattern of events following section of a motor nerve in *Periplaneta americana* outlined by Bodenstein (1957) has recently been re-examined by Guthrie (1964, 1967) and by Jacklet and Cohen (1967). The degree of regeneration of severed or crushed leg motor nerves is variable with respect to time course and extent, but a complete return to the original state is never achieved. Hyperplasia of regenerating axons may contribute to development of excessive muscle tension. Immature animals regenerate more extensively than adults but in the adult, age is not apparently a significant factor. After cutting, the stumps separate by at least 1 mm. The proximal stump may coil on itself and continue retraction for about a week, after which growth is evident. During the second and third weeks fibers sprout and send out tapered axon processes, some of which grow toward the distal stump, some following tracheae. The distal stump remains in position and the cut motor axons degenerate within five days (Jacklet and Cohen, 1967) though fibers can be recognized in the distal stump as long as nine weeks after operation (Guthrie, 1967). The mechanism of bridge formation between proximal and distal stumps remains an intriguing question: whether the gap is first bridged by regenerating sensory fibers which provide a pathway for motor fibers is not yet known; such a pathway is not essential for regenerating motor fibers to reach a muscle surface. Regeneration of motor nerves by fusion of severed axons is postulated in Crustacea (Hoy *et al.*, 1967) but has not been reported in insects as yet.

The primary regeneration of denervated muscle in *Periplaneta* is achieved by a few widely distributed axons (Guthrie, 1967) and may occur within 30 days (Jacklet and Cohen, 1967b). Electrical stimulation of regenerating nerves first elicits irregular slow fiber responses, possibly because of small end plate dimensions in pioneer fibers. Weak, prolonged fiber responses change as reinnervation proceeds, giving an increase in amplitude and the rate of decline of contractions. During the early stages of regeneration, spontaneous fibrillation and tremors may be observed. Inhibitory fibers regenerate as do fast and slow fibers (Guthrie, 1967) but the temporal pattern in the establishment of functional connections is yet to be described. The growth and maintenance of muscle tissue requires motor innervation (Nuesch, 1968); neuron cell bodies in *Periplaneta* atrophy if no motor connections are established with muscle by 150 days after injury (Jacklet and Cohen, 1967) and in *Calliphora erythrocephala* the extirpation of mesothoracic limb rudiments in the late larvae causes a 37% reduction in mesothoracic neuropile volume of the adult (Chiarodo, 1963). These, and many similar observations are evidence for a reciprocal trophic relationship between motor nerve and muscle, as in Vertebrates (Singer, 1965).

The cricket *Acheta domesticus* resembles the cockroach in its capacity for regenerating peripheral fibers (Edwards, 1967b, and unpublished data) but the locust *Locusta migratoria* evidently does not (Usherwood, 1963). Nor has regeneration of connections between severed ventral nerve cords of *Galleria mellonella* been observed (Edwards, unpublished).

The response of the motor cell bodies to nerve section is prompt (Cohen and Jacklet, 1965). An increase in perinuclear basiphilia is detectable twelve hours after section, and increases up to three days after injury, declining thereafter to normal during two weeks. Increased incorporation of tritiated uridine during the formation of the perinuclear basiphilic zone indicated an acceleration of RNA synthesis. Evidently the initial response is the synthesis of Nissl-like material, that is, aggregated ribosomal material, for up to three days, followed by dispersion at a time coincident with the inception of regeneration, and thus comparable with the chromatolytic response of the vertebrate neuron (Jacklet and Cohen, 1967b). The nucleus moves to an asymmetric position in the cell at the inception of regeneration, and may remain in that position as long as 100 days after nerve section, until reinnervation of muscle tissue is complete. The insect neuron thus differs significantly from that of the Vertebrate since, in the absence of Nissl substance, ribosomal material must be synthesized and dispersed before protein synthesis and nerve repair may proceed.

B. NEURON GROWTH IN ISOLATED GANGLIA

Transplanted ganglia become tracheated and survive indefinitely (e.g. Schrader, 1938; Bodenstein, 1957). Cut surfaces of central nervous tissue are covered over by hemocytes within a few hours. Interneurons (Drescher, 1960) and motor neurons (Bodenstein, 1957; Guthrie, 1962, 1964, 1967; Jacklet and Cohen, 1967) will emerge and develop from cut surfaces. The optic and olfactory commissures can be reformed within 2–3 weeks in the brain of *Periplaneta* after section in the mid plane (Drescher, 1960). Glial cells may enlarge and migrate to regions where axon degeneration is occurring (Hess, 1958; Drescher, 1960; Guthrie, 1967).

Working with *Periplaneta americana,* Guthrie (1964) cut the fifth nerve after implanting thoracic or abdominal ganglia. After five weeks, the limb recovered activity, differing from normal only in the occurrence of hypercontraction of the tibial depressor which may be due to duplicated innervation. Considerable rearrangement occurred within the ganglion; sometimes transverse tracts were lost, and in some cases cell bodies migrated into the neuropile. Severing connections between donor abdominal ganglia and host thoracic ganglion after several weeks abolished contraction in muscles supplied by the fifth nerve: the ganglion evidently acted simply as a conduit for the reinnervation of the muscle from the host ganglion. When the central connections of the host ganglion were cut, after an implanted thoracic ganglion had formed connections with leg muscles, limb function was retained.

C. SPECIFICITY IN REGENERATING NEURONS

The reconnection of severed olfactory and optic commissures in the cockroach brain has been mentioned above. That several connections between brain and optic lobes can also be regenerated with a quality that restores optomotor responses was shown by Drescher (1960) who has also demonstrated that afferent fibers from regenerating antennae of animals from which half the brain had been removed will cross to the olfactory center of the contralateral side. Similarly, afferent fibers from antennae produced as heteromorphic regenerates from damaged eyes in Crustacea (Maynard, 1965; Maynard and Cohen, 1965) and probably in insects (Edwards, 1967b) reach olfactory centers. All these observations, as well as Guthrie's (1964) and Jacklet and Cohen's (1967a,b) work mentioned above, imply a high degree of specificity in regeneration of sensory and interneuron connections, but the persistence of normal responses mediated by eyes of *Locusta* after 180° rotation and subsequent regeneration of connections with the optic lobe (Horridge, 1968) suggest a lack of specificity in the mechanism directing spatial arrangement of fibers in the optic lobe.

Abdominal Cerci of the house cricket *Acheta domesticus,* which regenerate after amputation, provide a model system for the study of neural regeneration. During this process afferent fibers from newly formed receptor hairs on the cerci enter the terminal ganglion where they re-establish synaptic contact with the giant fibers (Edwards, 1967b; Edwards and Sahota, 1968). A cercal regenerate will develop after transplantation to the stump of a coxa, and will grow at a rate comparable to that of regenerates *in situ.* Centripetal fibers from these transplanted cercal regenerates enter the thoracic ganglion of the segment to which they are grafted and, as shown in silver preparations of such ganglia, approach the giant fibers (Edwards and Sahota, 1968). Giant fiber spikes are elicited in the mesothoracic connectives when the transplanted cerci on the same segment are stimulated mechanically and the response remains intact after the animal is decapitated, the abdominal nerve cord severed, and other possible extraneous sources of impulses eliminated by coating all the body but the grafted cercus with heavy oil. These results suggest that the centripetal cercal fibers recognize the cells of their central termination as a whole and not only the normal synaptic region of that cell.

The specificity evident in the regeneration of sensory fibers described above does not seem to be the rule in motor systems. Thoracic ganglia of the cockroach, transplanted into the coxa of a host, form connections with nearby musculature (Bodenstein, 1957; Guthrie, 1964, 1967; Jacklet and Cohen, 1967) but the innervation does not appear to establish specific connections with the musculature. Implanted abdominal ganglia appeared to serve merely as a conduit for regenerating fibers from the severed host ganglion, but implanted thoracic ganglia formed connections with denervated muscle along with the host ganglion. The implanted thoracic ganglia clearly establish synaptic contact with host muscle. Reflex responses were elicited after the host nerve was cut; thus some sensory connections seem to be established also (Guthrie, 1964; Jacklet and Cohen, 1967a,b). Only denervated muscle receives fibers.

Supernumerary legs transplanted to the mesothoracic pleura of *Acheta domesticus* regenerated after they were severed through the femur and were innervated from branches of the nearest leg nerve, N4b, which was damaged by the grafting operation. The regenerates developed innervated muscle tissue and made active movements, though no clear relationship to the walking movements of the normal limb could be detected in high-speed cine-film sequences (Sahota and Edwards, unpublished). Thus, innervation can proceed when the normal limb is intact, and branches of a severed nerve will innervate a supernumerary limb, but demonstration that innervation of "vacant" muscle tissue can be initiated when the host nervous system is intact is difficult, since the surgery required for grafting must disrupt some nerves.

D. THE ROLE OF THE NERVOUS SYSTEM IN
MORPHOGENESIS AND REGENERATION

The developing nervous system is itself controlling development of other tissues, and observations on this process relate to questions of specificity in the nervous system. Significant advances in knowledge of this function, due largely to Nuesch and students, have recently been reviewed (Nuesch, 1968) and it is sufficient here to observe that a detailed picture has emerged of the necessity for innervation of the developing musculature in metamorphosing Holometabola and in regenerating structures such as legs, in which the musculature is critical for function. In Hemimetabola too, despite the capacity for regeneration of motor fibers, growth in denervated muscles is suppressed after nerves are severed. Only muscle tissue is directly involved in this control: tracheae and epidermis are unaffected, although Schoeller (1964) has presented a case for the initiation of differentiation in the fly antennal disk by centrifugal fibers (see section V B), and Urvoy (1963) reports the transformation to an antennal structure of a foreleg transplanted to an antennal stump in *Blabera craniifer*. An alternative explanation in the latter case may be that the regenerate was "recaptured" by the antennal cells and that the nervous system thus played no inductive role. Grafts of regenerating cerci to stumps of mesothoracic legs in *Acheta* either yielded a leg regenerate, formed a mosaic showing both leg and cercal cuticle, or developed as a pure cercus (Edwards and Sahota, 1968). The vigor of regeneration of antennae is attested by their occasional development as heteromorphic regenerates after cautery of the compound eye in *Acheta* (Edwards, 1967b). Further, many examples of homoeotic regeneration of antennae and of homoeotic mutants are known (Wigglesworth, 1965) which are difficult to explain in terms of a specific influence of centrifugal fibers, and Suster's (1933) demonstration that antennae are regenerated in *Sphodromantis* independently of nerve supply argues against a specific role for centrifugal fibers in the determination of a regenerate. A role of central nervous tissue in the capacity for regeneration, though not in the determination of the regenerate, is nevertheless implied by

Drescher's (1960) study of the effects of brain ablation in *Periplaneta*. Two difficulties complicate the interpretation of these, and comparable studies not cited here (see Drescher, 1960). The first is the need for electron microscopy of tissues before statements about presence or absence of nerves can be made without reservation. The second is the complex relationship between neural and humoral control mechanisms. An example of the latter is the recent demonstration by Kunkel (1967) that the delay of molt in *Blattella* caused by leg amputation is mediated by nervous input to the central nervous system, probably from mechanoreceptors, which delays the inception of the molting cycle, rather than a humoral response as postulated by Penzlin (1964, 1965).

ACKNOWLEDGEMENTS

I am grateful to Dr. Hans Nuesch for the opportunity to read his manuscript on related topics (Nuesch, 1958). I have also benefited from Dr. R. Nordlander's knowledge of lepidopteran brain development. Grant NB 05137 from the National Institutes of Health enabled me to carry out work discussed above.

REFERENCES

Afify, A. M. (1960). Uber die postembryonale Entwicklung des Zentralnervensystems (ZNS) bei der Wanderheuschrecke *Locusta migratoria migratorioides* (R. U. F.) (Orthoptera-Acrididae). *Zool. Jb. Abt. Anat.* **78**, 1–38.

Altman, J. and Das, G. P. (1965). Post-natal origin of microneurones in the rat brain. *Nature, Lond.* **207**, 953–956.

Ashhurst, D. E. and Richards, A. Glenn. (1964). A study of the changes occurring in the connective tissue associated with the central nervous system during the pupal stage of the luna moth. *J. Morph.* **114**, 225–236.

Bauer, V. (1904). Zur Inneren Metamorphose des Zentralnervensystems der Insekten. *Zool. Jb. Abt. Anat.* **20**, 123–152.

Beier, M. (1928). Zur Zytologie des Nervensystems der Insekten wahrend der Metamorphose. *Zool. Anz.* **77**, 52–56.

Bierbrodt. (1942). Der Larvenkopf von *Panorpa communis* L. und seine verwandlung mit besonderer Berucksichtigung des Gehirns und der Augen. *Zool. Jb. Abt. Anat.* **68**, 49–136.

Bodenstein, D. (1953). Postembryonic Development. In *Insect Physiology* (K. D. Roeder, ed.) Wiley, pp. 822–865.

Bodenstein, D. (1957). Studies on nerve regeneration in *Periplaneta americana*. *J. exp. Zool.* **136**, 89–116.

Brandt, E. (1879). Uber das Nerven System der Laufkafer. (Carabidae). *Trudy USSR ent. Obshch.* **14**, 4–5.

Bretschneider, F. (1914). Uber die Gehirne des Goldkafers und des Leder-laufkafers. *Zool. Anz.* **43**, 490–497.

Bullock, T. H. and Horridge, G. A. (1965). In *Structure and Function in the Nervous Systems of Invertebrates*. W. H. Freeman and Co., San Francisco.

Burtt, E. T. and Catton, W. Y. (1966). Image formation and sensory transmission in the compound eye. In *Advances in Insect Physiology* (J. W. L. Beament, J. E. Treherne and W. B. Wigglesworth, eds.) Vol. 3, pp. 1–52. Academic Press, London and New York.

Cajal, S. R. and Sanchez, D. (1921). Sobre le estructura de los centros opticos de los insectos. *Revta chil. Hist. nat.* **25**, 1–18.

Carlson, J. G. (1961). The grasshopper neuroblast culture technique and its value in radiobiological studies. *Ann. N.Y. Acad. Sci.* **95**, 932–941.

Carlson, J. G. and Gaulden, M. E. (1964). Grasshopper neuroblast techniques. In *Methods in Cell Physiology*. Vol. 1, pp. 229. (D. M. Prescott, ed.). Academic Press, New York.

Chiarodo, A. T. (1963). The effects of mesothoracic leg disc extirpation on the postembryonic development of the nervous system of the blowfly *Sarcophaga bullata. J. exp. Zool.* **153**, 263–277.

Cloarec, A. and Gouranton, J. (1965). Contribution a l'etude de l'organogenese des centres nerveux protocerebraux du criquet *Locusta migratoria migratorioides. Bull. biol. Fr. Belg.* **99**, 357–368.

Cody, F. P. and Gray, I. E. (1938). The changes in the central nervous system during the life history of the beetle *Passalus cornutus* Fabricius. *J. Morph.* **62**, 503–517.

Cohen, M. J. and Jacklet, J. W. (1965). Neurones of insects: RNA changes during injury and regeneration. *Science, N.Y.* **48**, 1237.

deLong, G. R. and Sidman, R. L. (1962). Effects of eye removal at birth on histogenesis of the mouse superior colliculus: an autoradiographic analysis with tritiated thymidine. *J. comp. Neurol.* **118**, 205–224.

Drescher, W. (1960). Regenerationsversuche am Gehirn von *Periplaneta americana. Z. Morph. Okol. Tiere,* **48**, 576–649.

Edwards, J. S. (1967a). Neural control of development in Arthropods. In *Invertebrate Nervous Systems* (C. A. G. Wiersma, ed.) pp. 95–110. University of Chicago Press.

Edwards, J. S. (1967b). Some questions for the insect nervous system. In *Insects and Physiology* (J. W. L. Beament and J. E. Treherne, eds.) pp. 163–173. Oliver and Boyd, Edinburgh.

Edwards, J. S. and Sahota, T. S. (1968). Regeneration of a sensory system: the formation of central connections by normal and transplanted cerci of the house cricket *Acheta domesticus. J. exp. Zool.* **166**, 387–396.

El Shatoury, H. H. (1956). Differentiation and metamorphosis of the imaginal optic glomeruli of *Drosophila. J. Embryol. exp. Morph.* **4**, 240–247.

Finlayson, L. H. (1956). Normal and induced degeneration of abdominal muscles during metamorphosis in the Lepidoptera. *Q. Jl. microsc. Sci.* **97**, 215–233.

Finlayson, L. H. (1960). A comparative study of the effects of denervation on the abdominal muscles of saturniid moths during pupation. *J. Insect Physiol.* **5**, 108–119.

Gieryng, R. (1965). Veranderungen der histologischen Struktur des Gehirns von *Calliphora vomitoria* (L.) (Diptera) wahrend der postembryonalen Ent-wicklung. *Z. wiss. Zool.* **171**, 80–96.

Graichen, E. (1936). Das Zentralnervensystem von *Nepa cinerea* mit Einschluss des sympatischen Nervensystems. *Zool. Jb. Abt. Anat.* **61**, 195–238.

Guthrie, D. M. (1964). Physiological competition between host and implanted ganglia in an insect (*Periplaneta americana*). *Nature, Lond.* **210**, 312–313.

Guthrie, D. M. (1967). The regeneration of motor axons in an insect. *J. Insect. Physiol.* **13**, 1593–1611.

Gymer, A. and Edwards, J. S. (1967). The development of the insect nervous system. I. An analysis of postembryonic growth in the terminal ganglion of *Acheta domesticus. J. Morph.* **123**, 191–197.

Haas, G. (1956). Entwicklung des Komplexauges bei *Culex* und *Aedes aegypti. Z. Morph. Okol. Tiere,* **45**, 198–216.

Hamburger, V. and Levi-Montalcini, R. (1949). Proliferation and degeneration in the spinal ganglia of the chick embryo under normal and experimental conditions. *J. exp. Zool.* **111**, 457–502.

Hanstrom, B. (1925). Comparison between the brains of the newly hatched larva and the imago of *Pieris brassicae. Ent. Tidskr.* **46**, 43–52.

Hanstrom, B. (1926a). Untersuchungen uber die relative Grosse Gehirnzentren verschiedener Arthropoden unter Berucksichtigung der Lebensweise. *Z. mikrosk-anat. Forsch.* **7**, 135–190.

Hanstrom, B. (1926b). Das Nervensystem und die Sinnesorgane von *Limulus polyphemus. Acta Univ. lund: Avd.* **2**, **22(5)**, 1–79.

Hanstrom, B. (1940). Inkretorische Organe, Sinnesorgane und Nervensystem des Kopfes einiger neideren Insektenordnungen. *K. svenska Vetensk-Akad. Handl.* **(3)18(8)**, 1–266.

Heller, R. and Edwards, J. S. (1968). Regeneration of the compound eye in *Acheta domesticus* (Abstract). *Am. Zool.* **8**, 786.

Henneguy, L. F. (1903). Sur la multiplication des cellules ganglionaires dans les centres nerveux chez les Insectes a l'etat des larvae et des nymphes. *Bull. Soc. ent. Fr.* **1903**, 324–326.

Hess, A. (1958). Experimental anatomical studies of pathways in the severed central nerve cord of the cockroach. *J. Morph.* **103**, 479–502.

Heymons, R. (1895). *Die Embryonalentwicklung von Dermapteren und Orthopteren unter besondere Berucksichtigung der Keimblatterbildung.* Fischer, Jena, 136 pp.

Heywood, R. B. (1965). Changes occurring in the central nervous system of *Pieris brassicae,* L. (Lepidoptera) during metamorphosis. *J. Insect Physiol.* **11**, 413–430.

Hinke, W. (1961). Das relative postembryonale Wachstum der Hirnteile von *Culex pipiens, Drosophila melanogaster* und *Drosophilamutanten. Z. Morph. Okol. Tiere,* **50**, 81–118.

Horridge, G. A. (1968). Affinity of neurones in development. *Nature, Lond.* **219**, 737–740.

Hoy, R. R. Bittner, A. D. and Kennedy, D. (1967). Regeneration in crustacean motorneurons: evidence for axonal fusion. *Science, N.Y.* **156**, 251–252.

Hughes, A. F. W. (1959). Studies in embryonic and larval development in Amphibia. II. The spinal motor root. *J. Embryol. exp. Morph.* **7**, 128–145.

Hughes, A. F. W. (1961). Cell degeneration in the larval ventral horn of *Xenopus laevis. J. Embryol. exp. Morph.* **9**, 269–284.

Hughes, A. F. W. (1968). *Aspects of Neural Ontogeny.* Logos Press, London.

Illies, J. (1962). Das abdominale Zentralnervensystem der Insekten und seine Bedeutung fur Phylogenie und Systematik der Plecopteren. *Dt. ent. Tgsber.* **45**, 139–152.

Imberski, R. B. (1967). The effect of 5-fluorouracil on the development of the adult eye in *Ephestia kuhniella. J. exp. Zool.* **166**, 151–162.

Jacklet, J. W. and Cohen, M. J. (1967a). Synaptic connections between a transplanted Insect ganglion and muscles of the host. *Science, N.Y.* **156**, 1638–1640.

Jacklet, J. W. and Cohen, M. J. (1967b). Nerve regeneration: correlation of electrical, histological and behavioral events. *Science, N.Y.* **156**, 1640–1643.

Jacobson, M. (1966). Starting points of research in the ontogeny of behavior. In *Major Problems in Developmental Biology.* (M. Locke, ed.). Academic Press, New York.

Jawlowski, H. (1936). Uber den Gehirnbau der Kafer. *Z. Morph. Okol. Tiere,* **32**, 67–91.

Johannsen, O. A. and Butt, F. H. (1941). *Embryology of Insects and Myriapods.* McGraw-Hill, New York.

Johannson, A. S. (1957). The nervous system of the milkweed bug, *Oncopeltus fasciatus* (Dallas) (Heteroptera lygaeidae). *Trans. Am. ent. Soc.* **83**, 119–183.

Johansen, H. (1892). Die Entwicklung des Imagoauges von *Vanessa urticae,* L. *Zool. Jb. Abt. Anat.* **6**, 445–480.

Kaji, S. (1960). Experimental studies on the developmental mechanism of the Bar eye in *Drosophila melanogaster VI. Mem. Konan Univ. Sci. Ser.* **4**, 1–16.

Kennedy, J. S. (1967). Behaviour as physiology. In *Insects and Physiology.* (J. W. L. Beament and J. E. Treherne, eds.) Oliver and Boyd, Edinburgh.

Kopec, S. (1922). Mutual relationship in the development of the brain and eyes of Lepidoptera. *J. exp. Zool.* **36**, 459–467.

Korr, H. (1968). Postembryonales Hirnwachstum von *Orchesella villosa. Z. Morph.* **68**, 389–422.

Kunkel, J. G. (1967). Controls of development in cockroaches. Ph.D. Thesis, Western Reserve University.

Levi-Montalcini, R. (1963). Growth and differentiation in the nervous system. In *The Nature of Biological Diversity* (J. M. Allen, ed.) pp. 261–297. McGraw-Hill, New York.

Lockshin, R. A. and Williams, C. M. (1965). Programmed cell death I–V. *J. Insect Physiol.* **11**, 123–133, 601–610, 803–809, 831–844.

Lucht-Bertram, E. (1962). Das postembryonile Wachstum von Hirneteilen bei *Apis mellifica* L. and *Myrmeleon europaeus* L. *Z. Morph. Okol. Tiere,* **50**, 543–575.

Lyonet, P. (1762). *Traite anatomique de la chenille qui range le Bois de Saule.* Gosse and Pinet, La Haye.

Malzacher, P. (1968). Die Embryogenese des Gehirns paurometaboler Insekten Untersuchungen an *Carausius morosus* und *Periplaneta americana. Z. Morph. Okol. Tiere,* **62**, 103–161.

Maynard, D. M. (1965). The occurrence and functional characteristics of heteromorph antennules in an experimental population of spiny lobsters *Panulirus argus. J. exp. Biol.* **43**, 79–106.

Maynard, D. M. and Cohen, M. J. (1965). The function of a heteromorph antennule in a spiny lobster, *Panulirus argus*. *J. exp. Biol.* **43**, 55–78.

Menees, J. H. (1961). Changes in the morphology of the ventral nerve cord during the life history of *Amphimallon majalis* Razoumowskiv (Coleoptera Scarabeidae). *Ann. Anat. Soc. Am.* **54**, 660–663.

Murray, F. V. and Tiegs, O. W. (1935). The metamorphosis of *Calandra oryzae*. *Q. Jl. microsc. Sci.* **77**, 405–495.

Neder, R. (1959). Allometrisches Wachstum von Hirnteilen bei drei verscheiden grossen Scharbenarten. *Zool. Jb. Abt. allg. Zool. Physiol.* **77**, 411–464.

Newport, G. (1832). On the nervous system of the *Sphinx ligustri*, Linn. and on the changes which it undergoes during a part of the metamorphosis of the insect. *Phil. Trans. R. Soc.* **122**, 383–398.

Nordlander, R. H. and Edwards, J. S. (1968a). Morphological cell death in the post-embryonic development of the insect optic lobes. *Nature, Lond.* **218**, 780–781.

Nordlander, R. H. and Edwards, J. S. (1968b). Morphology of the larval and adult brains of the Monarch butterfly *Danaus plexippus plexippus* L. *J. Morph.* **126**, 67–94.

Nordlander, R. H. and Edwards, J. S. (1969a). Postembryonic brain development in the Monarch butterfly *Danaus plexippus plexippus* L. I. Cellular events. *Roux Arch. Entwicklungsmech Organ.* (In press).

Nordlander, R. H. and Edwards, J. S. (1969b). Postembryonic brain development in the Monarch butterfly *Danaus plexippus plexippus* L. II. Development of the optic lobes. *Roux. Arch. Entwicklungsmech Organ.* (In press).

Nordlander, R. H. and Edwards, J. S. (1969c). Postembryonic brain development in the Monarch butterfly *Danaus plexippus plexippus* L. III. Development of brain centers other than the optic lobes. Manuscript in preparation.

Nuesch, H. (1968). The role of the nervous system in insect morphogenesis and regeneration. *A. Rev. Ent.* **13**, 27–44.

Panov, A. A. (1957). The structure of the brain in insects in successive stages of postembryonic development. *Ent. Obozr.* **36**, 269–284.

Panov, A. A. (1959). Structure of the insect brain at successive stages of post-embryonic development. II. The central body. *Ent. Obozr.* **38**, 276–284.

Panov, A. A. (1960a). The structure of the insect brain during successive stages of postembryonic development. III. Optic lobes. *Ent. Obozr.* **39**, 86–105.

Panov, A. A. (1960b). The character of reproduction of the neuroblasts, neurilemma, and neuroglial cells in the brain of the chinese oak silkworm larvae. *Dokl. Akad. Nauk SSSR*, **132**, 689–692.

Panov, A. A. (1961a). The structure of the insect brain at successive stages of postembryonic development. IV. The olfactory center. *Ent. Obozr.* **40**, 259–271.

Panov, A. A. (1961b). Ontogenetische Entwicklung des Zentralnervensystems von *Antheraea pernyi* Guer. (Lepidoptera). *Zool. Anz.*, **167**, 241–245.

Panov, A. A. (1962). The nature of cell reproduction in the central nervous system of the nymph of the house cricket (*G. domesticus* L., Orthoptera, Insecta). *Dokl. Akad. Nauk SSSR*, **143**, 471–474.

Panov, A. A. (1963). The origin and fate of neuroblasts, neurones, and neuroglial cells in the central nervous system of the Chinese silkmoth, *Antheraea pernyi* Guer (Lepidoptera, Attacidae). *Ent. Obozr.* **42**, 337–350.

Panov, A. A. (1966). Correlations in the ontogenetic development of the central nervous system in the house cricket *Gryllus domesticus* L. and the mole cricket *Gryllotalpa gryllotalpa* (Orthoptera, Grylloidea). *Ent. Obozr.* **45**, 179–185.

Penzlin, H. (1964). Die Bedeutung des Nervensystems fur die Regeneration bei den Insekten. *Roux Arch. Entwicklungsmech Organ.* **155**, 152–161.

Penzlin, H. (1965). Die Bedeutung von hormonen fur die Regeneration bei Insekten. *Zool. Jb. Abt. allg. Zool. Physiol.* **71**, 584–594.

Pflugfelder, O. (1958). *Entwicklungsphysiologie der Insekten.* Akademische verlagsgesellschaft, Leipzig.

Pipa, R. L. (1963). Studies on the hexapod nervous system. VI. Ventral nerve cord shortening; a metamorphic process in *Galleria mellonella* (L.) (Lepidoptera, Pyrallidae). *Biol. Bull. mar. biol. Lab., Woods Hole,* **124**, 293–302.

Pipa, R. L. (1967). Insect neurometamorphosis. III. Nerve cord shortening in a moth, *Galleria mellonella* (L.) may be accomplished by humoral potentiation of neuroglial motility. *J. exp. Zool.* **164**, 47–60.

Pipa, R. L. and Woolever, P. S. (1964). Insect neurometamorphosis. I. Histological changes during ventral nerve cord shortening in *Galleria mellonella* (L.) (Lepidoptera). *Z. Zellforsch. mikrosk. Anat.* **63**, 405–417.

Pipa, R. L. and Woolever, P. S. (1965). Insect neurometamorphosis. II. The fine structure of perineurial connective tissue. Adipohemocytes and the shortening ventral nerve cord of a moth, *Galleria mellonella. Z. Zellforsch. mikrosk. Anat.* **68**, 80–101.

Poulson, D. F. (1956). Histogenesis, organogenesis and differentiation in the embryo of *Drosophila melanogaster* Meigen. In *Biology of Drosophila melanogaster* (M. Demerec, ed.) pp. 168–174. Wiley, New York.

Power, M. E. (1943). The effect of reduction in numbers of ommatidia upon the brain of *Drosophila melanogaster. J. exp. Zool.* **94**, 33–72.

Power, M. E. (1946). An experimental study of the neurogenetic relationship between optic and antennal sensory areas in the brain of *Drosophila melanogaster. J. exp. Zool.* **103**, 429–461.

Power, M. E. (1952). A quantitative study of the growth of the central nervous system of a holometabolous insect, *Drosophila melanogaster. J. Morph.* **91**, 389–411.

Rensch, B. (1959). Trends toward progress of brains and sense organs. *Cold Spring Harb. Symp. quant. Biol.* **24**, 291–303.

Richard, G. and Gaudin, G. (1959). La morphologie du development du systeme nerveux chez diverse Insectes. In *The Ontogeny of Insects.* Acta Symposii de Evolutione Insectorum Prague.

Risler, H. (1954) Die somatische Polyploidie in der Entwicklung der Honigbiene und die Wiederherstellung der Diploidie bei den Drohnen. *Z. Zellforsch. mikrosk. Anat.* **41**, 1–78.

Rockstein, M. (1950). The relation of cholinesterase activity to change in cell number with age in the brain of the adult worker honey bee. *J. cell comp. Physiol.* **35**, 11–23.

Roonwall, M. L. (1937). Studies on the embryology of the African migratory locust *Locusta migratoria migratorioides. Phil. Trans. R. Soc. (B),* **227**, 175–244.

Sahota, T. S. and Beckel, W. E. (1967a). Orientation determination of developing flight muscles in *Galleria mellonella. Can. J. Zool.* 45, 407–420.

Sahota, T. S. and Beckel, W. E. (1967b). The influence of epidermis on the developing flight muscles in *Galleria mellonella. Can. J. Zool.* 45, 421–434.

Sanchez, D. (1924). Influence de l'histolyse des centres nerveux des insectes sur les metamorphoses. *Trab. Lab. Invest. biol. Univ. Madr.* 21, 385–422.

Sanchez, D. (1925). L'histogenese dans les centres nerveux des insectes pendant les metamorphoses. *Trab. Lab. Invest. biol. Univ. Madr.* 23, 29–52.

Satija, R. C. and Kaur, S. (1966a). Post-embryonic development of the optic ganglia in *Bombyx mori. Res. Bull. Panjab Univ. Sci.* 17, 297–304.

Satija, R. C. and Kaur, S. (1966b). Post-embryonic development of the brain in *Bombyx mori. Res. Bull. Panjab Univ. Sci.* 17, 353–365.

Schoeller, J. (1964). Recherches descriptives et experimentales sur la cephalogenese de *Calliphora erythrocephala* (Meigen). *Archs Zool. exp. gen.* 103, 1–216.

Schrader, K. (1938). Untersuchungen uber die Normalentwicklung des Gehirns und Gehirntransplantationen bei der Mehlmotte *Ephestia kuhniella* Zeller. *Biol. Zbl.* 58, 52–90.

Sehnal, F. (1965). Einfluss des Juvenilhormons auf die metamorphose des Oberschlund-ganglions bei *Galleria mellonella* L. *Zool. Jb. Abt. Physiol.* 71, 659–664.

Shafiq, S. A. (1954). A study of the embryonic development of the gooseberry sawfly *Pteronidea ribesii. Q. Jl. microsc. Sci.* 95, 93–114.

Sidman, R. L. (1963). Organ-culture analysis of inherited retinal degeneration in rodents. *Natn. Cancer Inst. Monogr.* 11, 227.

Singer, M. (1965). A theory of the trophic nervous control of amphibian limb regeneration. In *Regeneration in Animals and Related Problems* (V. Kiortsis and H. A. L. Trampusch, eds.) pp. 20–32. Amsterdam.

Sperry, R. W. (1965). Embryogenesis in behavioral nerve nets. In *Organogenesis* (R. L. Dehaan and H. Ursprung, eds.) pp. 161–186. Holt, New York.

Springer, C. A. (1967). Embryology of the thoracic and abdominal ganglia of the large milkweed bug *Oncopeltus fasciatus. J. Morph.* 122, 1–18.

Suster, P. M. (1933). Fuhlerregeneration nach ganglion extirpation bei *Sphodromantis bioculata* Burm. *Zool. Jb. Abt. allg. Zool. Physiol.* 53, 41–48.

Szekely, G. (1966). Embryonic determination of neural connections. In *Advances in Morphogenesis* (M. Abercrombie and J. Brachet, eds.) Vol. 5, pp. 181–219. Academic Press, New York.

Sztern, H. (1914). Wachstumsmessungen an *Sphodromantis bioculata* Burm. *Roux Arch. Entwicklungsmech Organ.* 40, 429–495.

Titschack, E. (1928). Der Fuhlernerv der Bettwanze *Cimex lectularius* L. und sein zentrales Endgebeit. *Zool. Jb. Abt. allg. Zool.* 45, 437–462.

Trager, W. (1937). Cell size in relation to the growth and metamorphosis of the mosquito, *Aedes aegypti. J. exp. Zool.* 76, 467–489.

Ullmann, S. (1967). The developing nervous system and other ectodermal derivatives in *Tenebrio molitor* L. (Insecta, Coleoptera). *Phil. Trans. R. Soc. (B),* 252, 1–25.

Umbach, W. (1934). Entwicklung und Bau des Komplexauges der Mehlmotte, *Ephestia kuhniella* Zeller, enbst einigen Bemerkungen uber die Entstehung der optischen Ganglien. *Z. Morph. Okol. Tiere*, **28**, 561–594.

Urvoy, J. (1963). Etude anatomo-functionelle de la patte et de l'antenne de la blatte *Blabera craniifer* Burmeister. *Annls Sci. nat.* **12**, 287–414.

Usherwood, P. N. R. (1963). Responses of insect muscle to denervation. II. Changes in neuromuscular transmission. *J. Insect Physiol.* **9**, 811–825.

Viallanes, M. H. (1882). Recherches sur l'histologie des insects et sur les phenomenes histologique qui accompagnent le development postembryonnaire de ces animaux. *Annls Sci. nat.* **14**, 1–348.

Wagner, H. G. and Wolbarsht, M. L. (1963). Electrical responses from transplanted insect eyes. *Fedn Proc. Fedn Am. Socs exp. Biol.* **22**, 519.

Weismann, A. (1864). Die nachembryonale Entwicklung der Musciden nach Beobachtungen an *Musca vomitoria* und *Sarcophaga carnaria. Z. wiss. Zool.* **14**, 187–336.

White, R. H. (1961). Analysis of the development of the compound eye in the mosquito, *Aedes aegypti. J. exp. Zool.* **148**, 223–240.

White, R. H. and Sundeen, C. D. (1967). The effect of light and light deprivation upon the ultrastructure of the larval mosquito eye. I. Polyribosomes and endoplasmic reticulum. *J. exp. Zool.* **164**, 461–477.

Wigglesworth, V. B. (1942). The significance of "chromatic droplets" in the growth of insects. *Q. Jl. microsc. Sci.* **83**, 141–152.

Wigglesworth, V. B. (1959). The histology of the nervous system of an insect *Rhodnius prolixus.* (Hemiptera). II. The central ganglia. *Q. Jl. microsc. Sci.* **100**, 299–313.

Wigglesworth, V. B. (1965). *The Principles of Insect Physiology.* 6th Edition. Methuen, London.

Witthoft, W. (1967). Absolute Anzahl und Verteilung der Zellen im Hirn der Honigbeine. *Z. Morph. Okol. Tiere,* **61**, 160–184.

Wolbarsht, M. L., Wagner, H. G. and Bodenstein, D. (1966). Origin of electrical responses in the eye of *Periplaneta americana.* In *The Functional Organization of the Compound Eye.* Wenner Gren Symposium 7. (C. G. Bernhard, ed.) pp. 207–217. Pergamon Press.

Wolsky, A. (1938). Experimentelle Untersuchungen uber die Differenzierung der zusammengesetzten Augen des Seidenspinners (*Bombyx mori* L.) *Roux Arch. Entwicklungsmech Organ.* **138**, 335–344.

Wolsky, A. (1956). The analysis of eye development in insects. *Trans. N.Y. Acad. Sci.* **18**, 592–596.

The Biology of Pteridines in Insects

IRMGARD ZIEGLER

Botanical Institute, Darmstadt Institute of Technology, Darmstadt, Germany

and

RUDOLF HARMSEN

Biology Department, Queen's University, Kingston, Canada

I. INTRODUCTION

The interest and research in a particular group of natural substances often originates with a few highly coloured representatives, and subsequently shifts to less obvious members of equal, or sometimes even greater, biological interest and importance. Groups such as flavines, carotenoides, porphyrins and purines are a few examples. The orange, yellow and white substances in the wings of pierid butterflies were studied in the nineteenth century (Hopkins, 1895). Later, these wing pigments were named "pterines" (πτερον-wing) by Wieland and Schöpf (1925). The term "pteridine" was introduced later (Schöpf and Becker, 1936); this term was confined to a chemically recognizable group of substances mainly responsible for wing pigmentation in Pieridae. Other groups of chemical substances also take part in the total pigmentation pattern: ommochromes, melanins, flavines, purines and various lipid pigments. The chemical composition of the pteridines became known a few years later when the structure of leucopterin was discovered (Purrmann, 1940; Schöpf and Reichert, 1941).

The long and arduous road which started with Hopkin's experiments on what he considered to be purines in butterfly wings, and ended with the chemical definition of leucopterin has been reviewed by Albert (1954) and more recently by Schöpf (1964). Our knowledge of the distribution and biological function of the pteridines is now extensive (Pfleiderer, 1963; Ziegler, 1965). These substances have been found in micro-organisms as well as plants and animals; in fact, all living organisms seem to contain at least small quantities of some pteridines. Insects appear to contain a higher concentration of total pteridine than most organisms (except for certain amphibians and reptiles) and also a larger variety of pteridines. Indeed, there are strong indications that the insects have developed more physiological functions, and thus a more varied use for the pteridines, than have other organisms.

II. PROPERTIES AND CHARACTERISTICS OF PTERIDINES

Pteridines are defined as those substances that are derived from the basic pyrimidine-pyrazine ring. Since most of the natural pteridines are 2-amino-4-hydroxy derivatives, this group of substances has become known as "pterines". All pterines can be represented as follows:

Folic acid and its derivatives, with a p-amino-benzoyl-glutamic acid substitution on C-6, are called "conjugated" pterines. The main conjugated pterines have the following composition:

Pteroic acid

Pteroylglutamic acid = folic acid: X = OH
Folic acid polyglutamate: X = 1-6 further glutamic acid units

The conjugated pterines will not be dealt with in this review as they form a very large and separate research field; this research has recently been reviewed (Jaenicke and Kutzbach, 1963). All other pterines could be called "unconjugated". These unconjugated pterines can best be classified in the following way for a biologically oriented discussion:

(1) Those substances with a polyhydroxy sidechain substituted in position C-6, including the not completely known drosopterins, e.g.

2-Amino-4-hydroxy-6-(L-erythro-1,2-di-hydroxypropyl)-pteridine = Biopterin

$$R_1 = \underset{\underset{OH}{|}}{\overset{\overset{H}{|}}{C}} - \underset{\underset{OH}{|}}{\overset{\overset{H}{|}}{C}} - CH_3 \quad R_2 = -H$$

2-Amino-4-hydroxy-6-(D-erythro-1,2,3-trihydroxypropyl)-pteridine = Neopterin

$$R_1 = \underset{\underset{H}{|}}{\overset{\overset{OH}{|}}{C}} - \underset{\underset{H}{|}}{\overset{\overset{OH}{|}}{C}} - \overset{\overset{OH}{|}}{CH_2} \quad R_2 = -H$$

(2) Those substances with only a carboxyl, hydroxyl or similarly simple substitution in C-6 and/or C-7 position. These are known as "simple" pterines, e.g.

2-Amino-4-hydroxypteridine = Pterin	$R_1 = R_2 = $ —H	
Pterin-6-carbonic acid	$R_1 = $ —COOH	$R_2 = $ —H
6-Hydroxypterin = Xanthopterin	$R_1 = $ —OH	$R_2 = $ —H
6,7-Dihydroxy pterin = Leucopterin	$R_1 = R_2 = $ —OH	
7-Hydroxypterin = Isoxanthopterin	$R_1 = $ —H	$R_2 = $ —OH
Isoxanthopterin-6-carbonic acid	$R_1 = $ —COOH	$R_2 = $ —OH

(3) The C-7 substituted pterines, including coupled di-pterines, e.g.

Erythropterin	$R_1 = $ —OH	$R_2 = $ —CH=COH—COOH
Lepidopterin	$R_1 = $ —OH	$R_2 = $ —CH=CNH_2—COOH
Ekapterin	$R_1 = $ —OH	$R_2 = $ —CH_2—CHOH—COOH
Pterorhodin		

(4) (a) The dihydrogenated pterines, which can be represented as follows:

and include substances such as

2-Amino-4-hydroxy-7,8-dihydro-pteridine = 7,8-Dihydropterin	$R_1 = R_2 = $ —H
6-Lactoyl-7,8-dihydropterin = Sepia-pterin	$R_1 = $ —CO—CHOH—CH_3 $R_2 = $ —H
6-Propionyl-7,8-dihydropterin = Iso-sepiapterin	$R_1 = $ —CO—CH_2—CH_3 $R_2 = $ —H
Dihydroekapterin	$R_1 = $ —OH $R_2 = $ —CH_2—CHOH—COOH

(b) The tetrahydrogenated pterines, which can be represented as follows:

and include substances such as

Tetrahydrobiopterin

$$R_1 = -\overset{\overset{\displaystyle H}{|}}{\underset{\underset{\displaystyle OH}{|}}{C}}-\overset{\overset{\displaystyle H}{|}}{\underset{\underset{\displaystyle OH}{|}}{C}}-CH_3 \qquad R_2 = -H$$

6-Lactoyltetrahydropterin $\qquad R_1 = -CO-CHOH-CH_3 \quad R_2 = -H$

Furthermore, the derivatives of 2,4-dihydroxy pteridine (lumazine) are of biological significance. They can be represented by the following general formula:

and include substances such as

2,4-Dioxotetrahydropteridine = 2,4-Di-
hydroxypteridine = Lumazine $\qquad R_1 = R_2 = -H$

2,4,7-Trioxotetrahydropteridine =
2,4,7-Trihydroxypteridine = 7-Hydroxy- $\quad R_1 = -H \qquad R_2 = -OH$
lumazine

6,7-Dimethyl-8-ribityllumazine $\qquad R_1 = R_2 = -CH_3$
$\qquad$ N(8): Ribitylmoiety

6-Methyl-7-hydroxy-8-ribityllumazine $\qquad R_1 = -CH_3 \quad R_2 = -OH$
$\qquad$ N(8): Ribitylmoiety

Also riboflavine is a lumazine derivative:

Albert (1954) and Pfleiderer (1963) have covered all chemical aspects of synthesis, reactability and structural properties of pteridines. Those properties of biological importance can be summarized as follows:

(a) Pteridines are insoluble in organic solvents. Conjugated and substituted pterines tend to dissolve readily in watery solutions. Simple pterines dissolve easily at low or high pH values, but around pH 7 these substances tend to autochelate, rendering them virtually completely insoluble.

(b) Most pteridines are amphoteric in nature.

(c) Dissolved pteridines, not in association with protein carriers, fluoresce when irradiated with UV light of high wave length (365 mμ). The colourless pteridines absorb strongly between 340–370 mμ, and this absorption is extended into the visible light (up to 500 mμ) in those pteridines which appear yellow, orange or red (Viscontini and Möhlmann, 1959a, b; Viscontini, 1958; Viscontini, et al., 1957; Viscontini and Stierlin, 1962a, b). Tetrahydro pterines do not fluoresce (Ziegler, 1963a). For review of UV absorption spectra of pteridines, see Mason (1954).

(d) The simple pterines (group 2) are chemically stable under normal physiological conditions, while the substituted (groups 1 and 3) and conjugated pterines are unstable. Light catalysed oxydation of the polyhydroxy side chain at C-6 is a normal phenomenon.

(e) The hydrogenated pterines are extremely unstable, especially at pH 7. Tetrahydrobiopterin, for instance, is oxidized within seconds in the light (Fig. 1), but more slowly in the dark (Ziegler, 1963a).

III.   SEPARATION, IDENTIFICATION AND LOCALIZATION

The usual method of separation of pteridines is paper chromatography. A variety of solvent systems has been successfully used (Viscontini et al., 1956; Viscontini et al., 1957; Rembold and Buschmann, 1963a; Harmsen, 1966b; Watt, 1964). The photosensitivity of many of them makes the use of chromatography in the dark or red light essential.

Thin layer chromatography with either silicone or aluminium oxide is unsuitable for the separation of pteridines, while column chromatography on cellulose powder gives excellent results. For the purification of large sample sizes of particular pterines, paper columns (Viscontini and Stierlin, 1962a, b) as well as phospho-cellulose and

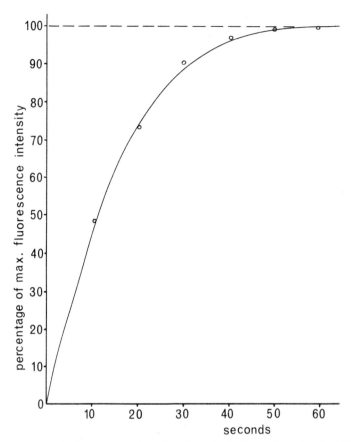

Fig. 1. Increase in fluorescence intensity of tetrahydrobiopterin under the influence of UV irradiation.

ECTEOLA cellulose (Rembold and Buschmann, 1962) have been used successfully. For the isolation of the extremely labile tetra-hydrobiopterin from the dihydro products, sephadex columns have been used with good results (Ziegler, 1963a; Dewey and Kidder, 1968). Paper electrophoresis has been used on a variety of pterine preparations especially for the drosopterins, sepiapterin and tetra-hydrobiopterin (Viscontini et al., 1957; Viscontini and Möhlmann, 1959a, b; Watt, 1964; Ziegler, 1960a).

The constant and extended linear relationship between fluorescence intensity of separated pterine samples on chromatography paper and the amount of substance in the sample makes a very useful tool for quantitative analysis. The sensitivity down to approximately $10 \mu g^{-2}$ allows very small sample sizes, which makes the use of individual insects, or parts of insects, feasible: a technique of particular importance in the study of mutants (Kühn, 1955; Ziegler and Hadorn, 1958). Several quantitative studies of pterine content in various tissues and at various developmental stages using fluorescence intensity measurements have been published (Harmsen, 1963, 1966c; Hudson et al., 1959; Watt, 1964, 1967).

The trypanosomid flagellate *Crithidia fasciculata* depends for its growth on the presence of a 2-amino-4-hydroxypteridine with a C-6 polyhydroxy side chain carrying at least two hydroxy groups in cis-arrangement. Biopterin (L-erythro-configuration) is the most active growth stimulating substance for *Crithidia* (Patterson et al., 1958). Biopterin and its derivatives can be analysed both qualitatively and quantitatively with the use of a *Crithidia* growth assay (the physiologically optimal concentration is approximately 0.3–0.5 mμg/ml). The detailed technique, and its various applications have been published by Nathan-Guttman (1964).

IV.   OCCURRENCE IN INSECTS

Practically all pteridines found in insects are 2-amino-4-hydroxy pteridines and thus pterines. The isoalloxazines, including riboflavine form the only obvious exception. Pterines are found in insects in a wide variety of substances: concentrated work on each new group of insects tends to reveal the occurrence of a new series of pterines (Schmidt and Viscontini, 1966; Harmsen, 1969b). Pterines are found in relatively large amounts in insects, and this is particularly true for the simple pterines which form the basis of pigmentation in certain Lepidoptera.

A.   PIGMENTS

When pterines with an absorption spectrum extending beyond 400 mμ are accumulated, their yellow, orange or red colour becomes apparent. General integument as well as eyes are often thus affected. Quantitative analyses of the wing pterines in Pieridae have revealed the concentration of these substances (Table I). Weygand and co-workers (1961) found in the wings of one specimen of *Pieris brassicae* approximately 500 μg leucopterin, while Harmsen (1966c) found approximately 350 μg. Watt (1967) found in the wings of one male of *Colias eurytheme* 197–268 μg xanthopterin, 98–136 μg leucopterin

Table I

Quantitative estimates and measurements of pterines in insects

	Pieris brassicae pharate adult (Harmsen, 1966c) (Weygand et al., 1961)	Vanessa io pharate adult (Harmsen, 1966a)	Colias eurytheme ♂ adult (Watt, 1967)	Apis mellifica pupa (Rembold and Bushman, 1963a, b) (Hanser and Rembold, 1964)	Oucopeltus fasciatus adult (Hudson et al., 1959)	Ceratitis capitata mature adult (Ziegler and Feron, 1965)
whole animal	470 µg leucopterin 85 µg isoxantho-pterin	37 µg leuc 17 µg isox	—	0.2 µg bio-pterin 0.015 µg neo-pterin	43 µg xanthopterin and erythropt 18.5 µg isox	—
eyes	—	—	—	—	—	1.1–1.2 µg tetrahydro-biopterin 1 µg sepia-pterine
body	120 µg leuc 25 µg isox	20 µg leuc 10 µg isox	98 µg leuc	—	—	—
wings	340 µg leuc (Harmsen) 500 µg leuc (Weygand) 55 µg isox	—	197–268 µg xan 98–136 µg leuc 265–53 µg erythr 19–28 µg sepiapt traces isox	—	—	—
meconium	10 µg leuc traces isox	—	—	—	—	—
food, queen	—	—	—	25 µg biopt/g 3 µg neopt/g	—	—
food, larva	—	—	—	4 µg biopt/g 0.3 µg neopt/g	—	—

Table II

Occurrence of pterines with pigmentation effect

Systematic orientation	Localization	Basis of Pigmentation	References
Hemiptera			
Dysdercus fasciatus	Entire integument of both adult and nymph	Erythropterin, unknown pterine related to erythropterin, xanthopterin and isoxanthopterin	Bartel *et al.*, 1958; Merlini and Mondelli, 1962; Merlini and Nasini, 1966; Harmsen, 1963, 1966a and unpublished data
Oncopeltus fasciatus			
Pyrrhocoris apterus			
Phonoctonus nigrofasciatus			
Other species			
Rhodnius prolixus	Entire integument of both adult and nymph	Erythropterin and lepidopterin	Viscontini and Schmidt, 1963
Cicadidae *(Rhynchotis)*	Wings	Xanthopterin	Becker, 1937b
Neuroptera			
Myremellontidae	Integument	Xanthopterin	Becker and Schöpf, 1936
Ascalaphidae	Wing spot	Xanthopterin	Becker, 1937b
Coleoptera			
Baldoria	Whole animals (Integument?)	Leucopterin and xanthopterin	Bernasconi, 1963
Speonomus			
Antrocharis			
Trechus			
Aphanops			
Hymenoptera			
Vespa spp.	Integument	Xanthopterin, leucopterin	Schöpf and Becker, 1933; Becker and Schöpf, 1936;

Ichneumonidae Chalcididae Mutillidae Scoliidae Tynnidae Pompilidae Sphegidae Apidae	Integument and wings	Xanthopterin	Becker, 1937b; Ikan and Ishay, 1967 Becker, 1937b
Lepidoptera *Bombyx mori*	Larval epidermis	Sepiapterin, 7,8-dihydro-6-lactoyl lumazine	Tsujita, 1961; Goto *et al.*, 1966
Vanessa io	Wing spot	Xanthopterin	Harmsen, 1963
Pieridae *Appias nero* *Appias drusilla* *Catopsilia argante* *Colias edusa* *Colias croceus* *Colias eurytheme* *Euchloe cardamines* *Gonepteryx rhamni* *Mylothris chloris* *Mylothris poppaea* *Phoebis argante* *Pieris napi* *Pieris brassicae*	Integumental scales and wings	Leucopterin isoxanthopterin, xanthopterin, erythropterin (often restricted to ♂ or in discoidal spots) sepiapterin (rarely, only in *C. eurytheme*)	Schöpf and Wieland, 1926; Schöpf and Becker, 1933, 1936; Becker and Schöpf, 1936; Becker, 1937b; Purrmann, 1940; Tschesche and Korte, 1951; Pfleiderer, 1962, 1963; Harmsen, 1963, 1966b, 1969a; Watt, 1964, 1967; Descimon, 1966; Watt and Bowden, 1966
Pieris brassicae	Larval integument	Leucopterin, sepiapterin	Harmsen, 1966b

and 25–50 μg erythropterin. The pterine concentration in the eyes of Diptera is also relatively high: from the eyes of one individual fly *(Ceratitis capitata)* were isolated approximately 1 μg of sepiapterin and 1 μg of tetrahydrobiopterin (Ziegler, 1963; Ziegler and Feron, 1965).

These pterines, which occur as pigments in the integument of Lepidoptera, Hemiptera and Hymenoptera, are mostly either simple pterines or C-7 substituted pterines, while the Dipteran eye colouring pterines are mainly C-6 substituted or hydrogenated pterines.

It is of interest to note that sepiapterin and isosepiapterin, although both 7,8-dihydro compounds, have a yellow appearance due to absorption above 400 mμ and fluoresce in UV light; the tetrahydropterines are neither coloured nor fluorescing. The chemical composition of the drosopterins is not completely clear (Forrest and Nawa, 1964; Viscontini, 1958, 1964). With paper chromatography they can be separated into three component substances: drosopterin, isodrosopterin and neodrosopterin (Viscontini *et al.*, 1957). All three substances are red in colour (absorption maximum 475–505 mμ) but discolour in the light within a few minutes when dissociated from their protein carrier. In this respect, neodrosopterin is the most labile (Viscontini, 1958; Ziegler and Nathan, 1961). Preliminary experiments concerning structure and biosynthesis (see p. 180) of the drosopterins show them to have a three carbon side chain in C-6 position, to be optically active, and probably to be di-hydrogenated. The very low Rf-values for most solvents on paper chromatography suggest that they are dipteridyl derivatives like pterorhodin. Growth experiments with *Crithidia fasciculata* (Ziegler and Nathan, 1961; Nathan and Ziegler, 1961) show that neodrosopterin is close to biopterin in its growth stimulating effect; drosopterin has only a 50% activity, while isodrosopterin is inactive. The 50% activity of drosopterin suggests that it is a dipteridyl compound consisting of a combination of opposing stereoisomers.

Tables II and III summarize the occurrence of pigment pterines in the integument and the eyes respectively.

In many cases it is difficult to assess whether a pterine functions as a pigment or not. Leucopterin and isoxanthopterin are responsible for the white appearance of the wings of *Pieris,* but the isoxanthopterin in the integument of the larva of *Pieris* or in the eyes of *Drosophila* carries no colouring effect. The sepiapterin in the wild type *Drosophila melanogaster* is completely masked by the drosopterins and the ommochromes. In the *sepia* mutant, and especially in

Table III

Occurrence of pterines with pigmentation effect in the compound eye

Systematic orientation	Details	References
Diptera		
Drosophila melanogaster (wild type)	Drosopterins (drosopterin, neodrosopterin, isodrosopterin) contribute to red pigmentation	Hadorn and Mitchell, 1951; Ziegler and Hadorn, 1958; Viscontini *et al.*, 1957
Drosophila melanogaster (mutant *sepia*)	Sepiapterin and isosepiapterin, causing yellow-brown pigmentation	Hadorn and Mitchell, 1951; Ziegler and Hadorn, 1958
Calliphora erythrocephala	Sepiapterin, masked by ommochromes in wild type, but giving yellow colour to mutant *W*	Ziegler, 1961a; Patat, 1965
Ceratitis capitata	Sepiapterin, masked by ommochromes	Ziegler, 1963a; Ziegler and Feron, 1965
Glossina pallidipes and other spp of *Glossina*	Drosopterins in minute quantities, sepiapterin and isosepiapterin, pterine PR_1 (unidentified) most abundant red pterine pigment	Harmsen, 1969b
Aedes aegypti, A. mascariensis, Culex pipiens	Sepiapterin	Bhalla, 1968
Lepidoptera		
Ephestia kühniella *Ptychopoda seriata*	Erythropterin and pterorhodin contribute to red eye pigmentation. Obvious in ommochrome free mutant *a* of *Ephestia*	Kühn and Egelhaaf 1959; Viscontini and Stierlin, 1962a, b

the combination *se/se; cn/cn,* where both drosopterins and ommo-chromes are missing, sepiapterin becomes the main eye pigment.

The original contention that certain insects are pterine containing and others are not (Becker, 1937a), was based on the presence or absence of accumulations of pigment pterines. Obviously, all insects contain pterines, but in only certain groups are pterines functioning as pigments.

B. COLOURLESS PTERINES

Pigment pterines are always found in association with colourless ones, the commonest of which are isoxanthopterin, leucopterin, 2-amino-4-hydroxypteridine and biopterin. In some cases coloured pteridines are present in quantities insufficient to play a role in pigmentation. The most generally distributed of this latter kind are sepiapterin and riboflavine. The concentrated occurrence of colour-less pterines, however, is not necessarily associated with the occur-rence of coloured ones, as in many cases they are found on their own (Table IV).

C. HYDROGENATED PTERINES

It is often possible to show that at least most of the fluorescing pterines, such as pterine-6-carbonic acid, biopterin and 2-amino-4-hydroxy pteridine itself, when found in tissue extracts, are degrada-tion products of tetrahydrobiopterin. This important substance does not fluoresce, but absorbs strongly at approximately 300 mμ. At pH values above 7.5, it becomes dehydrogenated relatively quickly during chromatography even in the dark, while in the light it becomes dehydrogenated within seconds (see Fig. 1). This rapid dehydrogena-tion, and often part degradation, is the cause of tetrahydrobiopterin being overlooked so often. Only extreme care during separation will result in a true assessment of its presence (Ziegler, 1960a, 1963a). There is a suspicion that the reported occurrence of the oxidized pteridines is really, in many cases, a cryptic indication of the presence of tetrahydrobiopterin. Pterine-6-carbonic acid may not be a naturally occurring substance at all. This situation may be true especially for those tissues where the oxidized pterines are not accumulated but have been found merely in low or very low concentrations. For instance, a surprising discovery was the presence of traces of tetrahydrobiopterin in the meconium of *Drosophila melanogaster* (Wessing and Eichelberg, 1968).

Careful examination of Sephadex analysed extracts of *Ceratitis capitata* has shown that the tetrahydrobiopterin is present in an

Table IV

Occurrence of pterines without pigmentation effect

Systematic orientation	Localization	Pterine	References
Orthoptera			
Locusta			
Schistocerca			
Schistocerca gregaria	Eyes	Xanthopterin resembling pterines	Goodwin, 1951
	Entire ♂ body	Xanthopterin and sepiapterin; the yellow pigmentation of the ♂ is not a pterine	Harmsen, 1966a
Locusta migratoria	Adult integument and eyes	Isoxanthopterin, xanthopterin, sepiapterin and other not identified fluorescing substances	Bouthier, 1966
Phasmida			
Carausius morosus	Entire body, especially nervous system	Xanthopterin, isoxanthopterin, 2-amino-4-hydroxypteridine, biopterin, pterine-6-carbonic acid	L'Hélias, 1961, 1962; Fischer *et al.*, 1962
Dictyoptera			
Periplaneta	Nervous system	Xanthopterin, isoxanthopterin, 2-amino-4-hydroxypteridine	Fischer *et al.*, 1962
Odonata			
Aeschna cyanea	Larval epidermis	Unidentified blue fluorescing pterine	Krieger, 1954
Hemiptera			
Sappaphis plantaginea			
Brevicoryne brassicae	Nervous system and other body parts	Xanthopterin, isoxanthopterin, biopterin, pterine-6-carbonic acid	L'Hélias, 1961, 1962

Table IV (cont.)

Systematic orientation	Localization	Pterine	References
Milletia utreldiana *Leptocoris apicalis* and nine other species	Entire adults	Isoxanthopterin probably chryso-pterin and violapterin, in association with integumental erythropterin	Merlini and Nasini, 1966
Dysdercus fasciatus *Phonoctonus nigro-fasciatus*	Integument and fatbody	Xanthopterin, leucopterin and isoxanthopterin, associated with erythropterin in the integument	Harmsen, 1963, 1966a; Bartel et al., 1958
Oncopeltus fasciatus *Oncopeltus fasciatus*	Entire adults and several body parts separately	Xanthopterin, isoxanthopterin, 2-amino-4-hydroxypteridine, biopterin, chrystopterin	Hudson et al., 1959; Forrest et al., 1966
Oncopeltus fasciatus	Eggs	Erythropterin, xanthopterin, 7,8-dihydroxanthopterin, isoxanthopterin (xanthopterin probably breakdown product of dihydroxanthopterin)	Hudson et al., 1959; Forrest et al., 1966
Coleoptera *Photinus*	Luminous organ	Luciferesceine (i.e. imino-ribityl-pterine)	Strehler, 1951
Luciola cruciata	Abdomen	Luciopterin (i.e. 8-methyl-trihydroxy pteridine)	Kishi et al., 1968
Tenebrio molitor	Entire body	Traces of isoxanthopterin and possibly biopterin	Harmsen, 1963
Hymenoptera *Mormoniella vitripennis*	Entire head	Unidentified red and yellow pterines	Saul, 1960
	Adult abdomen	Isoxanthopterin, 2-amino-4-hydroxypteridine and other, unidentified pterines	Saul, 1960

Species	Location	Pteridines	Reference
Apis mellifica	Entire pupae	Biopterin, 2-amino-4-hydroxy-pterine, isoxanthopterin, neopterin, violapterin, pterine-carbonic acid	Rembold and Buschman, 1963 a, b
	Queen food and worker	Biopterin, neopterin; both in much higher concentration in queen food (see Table I)	Rembold and Buschman, 1963a, b; Butenandt and Rembold, 1958
	Pharyngeal gland	Biopterin	Rembold and Hanser, 1960b
	Eyes and other internal organs	Isoxanthopterin, red fluorescing pterines and other fluorescing substances	Köhler, 1958
Formica rufa *Formica polyctena* *Formica cordieri*	Entire bodies of workers	2-Amino-4-hydroxy pteridine, isoxanthopterin, biopterin, formicapterin and riboflavine-formicapterin complex; pterines mainly in musculature, integument and gut	Schmidt and Viscontini, 1962, 1964
	Entire males	Isoxanthopterin, D- or L-6-(threo-1',2',3'-trihydroxypropyl) pterine, pterine carbonic acid; also isoalloxazine derivatives (lumazine, lumazine-6-carbonic acid, isoxantholumazine, 6-methyl-8-ribityl-isoxantholumazine)	Schmidt and Viscontini, 1966; Viscontini and Schmidt, 1966; Schmidt and Viscontini, 1967
Lepidoptera *Papilio xuthus* *Papilio protenor*	Pupal integument	Xanthopterin, leucopterin and isoxanthopterin both in green and brown pupae; does not effect pigmentation	Ohnishi, 1959
Vanessa io	Entire body and meconium	Leucopterin, isoxanthopterin, sepiapterin, xanthopterin	Harmsen, 1963, 1966a; Müller, 1956

Table IV (cont.)

Systematic orientation	Localization	Pterine	References
Pieris brassicae	Entire body of adult, pupa and larva	Isoxanthopterin, xanthopterin, leucopterin, sepiapterin (only in larva), biopterin, erythropterin	Harmsen, 1963, 1966a, b, c
Colias eurytheme	Meconium	Isoxanthopterin, leucopterin	Harmsen, 1966a
	Wings	Isoxanthopterin and sepiapterin in minute amounts	Watt, 1964, 1967
	Adult organs	Isoxanthopterin, leucopterin, sepiapterin	Watt, 1967
Appias nero	Larval organs	Leucopterin	Watt, 1967
Anthocaris cardamines	Wings	Isoxanthopterin and sepiapterin (with pigment pterines)	Pfleiderer, 1964
Colias croceus and 33 other species of Pieridae	Wings	Isoxanthopterin and sepiapterin (with pigment pterines)	Descimon, 1965a
Bombyx mori	Larval epidermis	Dihydroxanthopterin	Descimon, 1967
		Isoxanthopterin, leucopterin, 2-amino-4-hydroxy pteridine, biopterin, 7,8-dihydro-6-lactyl-lumazine	Tsujita and Sakaguchi, 1955; Goto *et al*, 1966
	Eggs	Blue and yellow fluorescing pterines	Polonovski and Busnel, 1948
Antheraea pernyi	Larval epidermis	Leucopterin, isoxanthopterin	Sakaguchi, 1955
Ephestia kühniella and *Plodia interpunctella*	Eyes	2-amino-4-hydroxy pteridine*, biopterin*, isoxanthopterin, xanthopterin, ekapterin, lepidopterin(?), other fluorescing substances; all these in much lower quantities than the pigment pterines: erythropterin and pterorhodin	Egelhaaf, 1956b; Viscontini *et al*, 1956; Almeida, 1958; Kühn and Almeida, 1961; Viscontini and Stierlin, 1962a, b; cf. Ziegler, 1961b

* Probably degradation products of hydrogenated pterines, especially tetrahydrobiopterin.

	Ovaries, eggs and testes	2-Amino-4-hydroxy pteridine, biopterin, xanthopterin, isoxanthopterin; lepidopterin (?) and others; also riboflavin	
	Vasa deferentia	Xanthopterin, isoxanthopterin, lepidopterin	
	Malpighian tubules, rectal sac, midgut and fat body	2-Amino-4-hydroxy pteridine, xanthopterin, isoxanthopterin, lepidopterin and others	Egelhaaf, 1956a
		2-Amino-4-hydroxypteridine, also several non-identified fluorescing substances, probably derivatives of tryptophane metabolism	
	Wings	2-Amino-4-hydroxy pteridine, biopterine, xanthopterin, isoxanthopterin and other unidentified fluorescing substances	
Diptera			
Phormia regina *Musca domestica* *Pollenia viridis* *Ceratitis capitata*	Eyes	Pterine-6-carbonic acid, biopterin, 2-amino-4-hydroxy pteridine (degradation products of tetrahydrobiopterin)	Ziegler, 1960a. 1963a
Glossina pallidipes and other species of *Glossina*	Eyes	Biopterin*, pteridine-6-carbonic acid*, 2-amino-4-hydroxy-pteridine, isoxanthopterin, xanthopterin and several unidentified fluorescing substances	Harmsen, 1969b
Calliphora erythrocephala	Eyes	Tetrahydrobiopterin, pterine-6-carbonic acid*	Ziegler, 1961a; Autrum and Langer, 1958

* Probably degradation products of hydrogenated pterines, especially tetrahydrobiopterin.

Table IV (cont.)

Systematic orientation	Localization	Pterine	References
Drosophila melanogaster	Eyes	2-Amino-4-hydroxy pteridine*, biopterin*, pterine-6-carbonic acid*, isoxanthopterin, xanthopterin and xanthopterin-like pterine (degradation product of sepiapterin), riboflavine; these occur with the drosopterins	Hadorn and Mitchell, 1951; Ziegler and Hadorn, 1958; cf. Ziegler, 1961b; Viscontini and Möhlmann, 1959a, b
	Malpighian tubules and testes	Isoxanthopterin, xanthopterin, sepiapterin, biopterin, 2-amino-4-hydroxy pteridine, riboflavin, and one unidentified fluorescing substance	Handschin, 1961; Hadorn *et al.*, 1958
	Meconium	Isoxanthopterin, 2-amino-4-hydroxy pteridine, biopterin, xanthopterin	Kürsteiner, 1961; Wessing and Eichelberg, 1968
Drosophila spp of two subgenera: *Sophophora* and *Drosophila*	Whole bodies (testes)	Drosopterins, sepiapterin, isoxanthopterin, xanthopterin, 2-amino-4-hydroxy pteridine, biopterin; the drosopterins present only in the more primitive representatives of both subgenera, especially in *virilis, quinaria, obscura* and *nannoptera*	Hubby and Throckmorten, 1960
Aedes aegypti, A. mascariensis, Culex pipiens	Eyes	Biopterin, 2-amino-4-hydroxy pteridine* accompanied by flavine-mononucleotide	Bhalla, 1968

* Probably degradation products of hydrogenated pterines, especially tetrahydrobiopterin.

unsubstituted form, and there was no indication of a pentose moiety (Ziegler, unpublished).

The obviously important role which the hydrogenated pterines play in the metabolism of the insect (see p. 170) more than warrants further research in the occurrence and localization of these substances.

Two di-hydrogenated pterines have been described (Forrest and Nawa, 1962): 6-lactoyl-7,8-dihydropterine (sepiapterin) and 6-propionyl-7,8-dihydropterine (isosepiapterin). These two dihydrogenated substances will also oxidize slowly in the dark, and faster in light, yet they are far less labile than tetrahydrobiopterin. The degradation products of the sepiapterins are mainly xanthopterin and a "xanthopterin-like" pterine (Ziegler and Hadorn, 1958). The recently described 7,8-dihydroxanthopterin and 7,8-dihydro-2-amino-4-hydroxy pteridine (Gutensohn, 1968) are probably di-hydrogenated intermediates in a degradative formation of simple pterines (see p. 182). Their presence in insects is highly probable (Harmsen, 1969a). Sepiapterin is equally as active a growth inducer for *Crithidia* as biopterin (Ziegler and Nathan, 1961).

Sepiapterin absorbs at approximately 440 mμ (Viscontini and Möhlmann, 1959a, b), which makes it possible to recognize its presence in tissues by its yellow colour, at least when other pigments are not present. When bound to protein granules it may appear dark brown (see p. 164).

Patat (1965) has suggested that tetrahydrobiopterin is also associated with protein granules in the tissues, and that this association is the cause of its relative photostability while in the eye. The photostability is maintained as long as the pigment is bound to protein in dead and liophylized eyes, as well.

Besides biopterin and related pterines there are non-fluorescing substances which have a *Crithidia* growth effect. These substances are not a tetrahydropterine, in that irradiation or oxydation do not produce fluorescing degradation products. Rembold and Hanser (1960b) have reported finding such substances in the gonads and in the larvae of the honey bee. In *Drosophila* they are present in larval homogenates as well as in the haemolymph of emerging adults (Ziegler, unpublished data). The pupa of *Pieris* also contains these *Crithidia* growth factors until the time of pterine deposition in the pharate adult cuticular scales. Chemical analysis (Schmidt, 1965) suggests that the main substance is a phosphorylated iminoribose substituted triamino-oxypyrimidine, a substance which seems to be close to the initial pterine in the purine $\rightarrow$ pterine transformation (cf. Fig. 6 and p. 178).

V. LOCALIZATION IN THE TISSUES

A. IN THE INTEGUMENT

The first detailed study on the localization of pterines in insect tissues was Becker's (1937b). This author found that the typical pigment pterines seem to be confined to those areas of the integument that do not border on tissues with a high metabolic turnover. In Pieridae and Vespidae the areas of the integument that carry the pterine concentrations are either skin folds (wings or body scales) or areas overlying tracheal sacs. In contrast to melanin, which is usually embedded inside the cuticle, the pterines are located in the epidermal cells. In the case of wing pigments, the pterines are found in the interlamellar space, associated with both the upper and lower cuticular plates. In Pieridae, however, most of the pterines are deposited in the minute interlamellar spaces of scales and hairs, mainly of the wings, but also of the body. In these positions the pterines are crystalline and show pleochroism, again differing from the diffusely deposited melanin pigments. In penetrating light the pterine crystals appear dark, while in reflecting light they take on a very bright appearance. As a result of the regular orientation of the optical axis of the crystals, the brightness varies with the angle of the light beam. It is not established whether the different wing pterines crystallize in mixed or separate crystals or granules, although it can be easily seen that in most Pieridae some scales seem to carry one pterine species almost exclusively, while other scales contain another pterine. Electron microscopy of the butterfly wing scale pterine granules shows them as $0.5-1.0 \mu \times 0.15-0.18 \mu$ elliptical granules, without internal structure fixed onto the cuticle (Descimon, 1965b). Other wing pigments (like ommochromes of the Nymphalidae and flavonoids of the Satyridae) are not granular, the scales being homogeneously pigmented.

Differing from the simple pterines, which are crystallized in the integument, sepiapterin, which forms the yellow pigment in the larval epidermis of the *lemon* mutant of *Bombyx mori*, is associated with protein granules (Tsujita and Sakurai, 1963). The granules are round, $1-2.5 \mu$ in diameter in the wild type, $2-3 \mu$ in *lemon*. Their appearance and chemical nature is similar to the pterine pigment granules of many insect eyes (see below).

B. IN THE COMPOUND EYE

In the ommatidia of the compound eye both the coloured pterines and the ommochromes are concentrated in the form of round protein-aceous granules in the pigment cells which surround the lens and retinula. Electron microscopy of these cells shows that they are remarkably poor in mitochondria, but rich in glycogen granules (Shoup, 1966; Langer, 1967). Most detailed work on the localization of pterines in the compound eye has been done in association with genetical studies *(Drosophila, Ephestia)* or electrophysiological studies *(Calliphora).* It is possible to combine the results of the cytological study of the distribution of pigment granules in the eyes of various *Drosophila* mutants (Zeutzschel, 1958) with the results of the eye pigment analysis of the same mutants (Ziegler, 1961a; cf. Ziegler, 1961b). Consequently, it is possible to determine the exact localiza-tion of the pterines (drosopterines and sepiapterin) and the ommo-chromes in the eye cells (Fig. 2), the pterines being found only in the

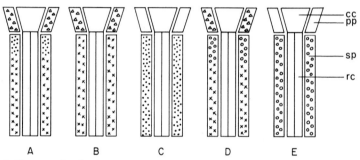

Fig. 2. Pigment distribution in the ommatidia of the compound eye of various mutants of *Drosophila melanogaster.* c.c. crystalline cone; p.p. primary pigment cell; s.p. secondary pigment cell; r.c. retinula cells. A. Wild type; B. *brown;* C. *vermilion* or *cinnabar;* D. *sepia;* E. *sepia/cinnabar.* △ ochre (ommochrome); X deep red (ommochrome); ● bright red (drosopterins); ○ yellow (sepiapterin). Redrawn after Zeutzschel (1958).

lateral pigment cells, where they are covered by ommochromes, mainly in the proximal end of the cell. Pterines are not present in the corneal pigment cells of the ommatidium, nor are they found in the ocelli. Only the separation from the protein granule and subsequent oxidation into a fluorescing degradation product will allow fluores-cent microscopical examination of tetrahydrobiopterin. However, this treatment results in the immediate spreading of the fluorescing material throughout the cell. Thus, it is only possible to conclude indirectly that tetrahydrobiopterin is found in the same cytological location as the other pterines. Only in those *Drosophila* mutants, where the coloured pterines are absent and the hydrogenated pterines have accumulated, is it possible to deduce that these hydrogenated substances are probably found in the same lateral pigment cells associated with the same brown pigment granules.

For the cornea pigment cells in *Ephestia kühniella* Hanser (1948) described the pigment as "xanthommatine". Histological and chromatographical analysis of a number of mutants (Reisener-Glasewald, 1956; Kühn and Berg, 1955) has shown that a yellow pterine pigment (possibly lepidopterin or ekapterin) is concentrated in these cells; the lateral and retinula pigment cells contain both ommochromes and the red pterines (pterorhodin and erythropterin). The *Ephestia* mutant *a*, which lacks all ommochromes, shows in the lateral and retinula pigment cells only bright red granules (Kühn and Berg, 1955). In *Ptychopoda seriata* the retinula cells contain no pigment. The lateral pigment cells are similar to those of *Ephestia*, while the description of the corneal pigment cells suggests that a pterine is also found here, in addition to the ommochromes. This possible pterine, called "skotommine", was not characterized at the time of Hanser's study (1948). Reisener-Glasewald's report (1956) of finding fluorescing pterines in the crystalline body is probably based on the diffusion of dissolved pterines at the time of tissue fixation (Carnoy's) and secondary absorption.

In all cases studied, with the possible exception of one (see p. 164), the typical eye pterines, as well as the ommochromes, occur bound to granules known as the pigment granules in the eye pigment cells. These granules usually have a 0.4–0.8 μ diameter. The only exception is isoxanthopterin which is also found in certain granule-free mutants, e.g. *Ephestia kühniella* mutant *wa/wa; bch/bch* (Kühn and Berg, 1956). It is also found in full concentration in the eyes of the pharate imago of *Drosophila*, before the appearance of the pigment granules (Hadorn and Ziegler, 1958), see also p. 176.

It has been established that the pigment granules consist not only of pigment, but also of a carrier substance. When certain histological techniques dissolve all the pigments, haematoxylin stainable granules of 0.2–0.3 μ diameter remain behind in the cell (Hanser, 1948–in *Ephestia).* In the ommochrome and pterine lacking mutant *wa*, these granules are absent. In the *Ephestia kühniella a* mutant and in the *Calliphora erythrocephala W* mutant, both lacking only the ommochromes, the carrier granules are present (Hanser, 1948, 1959). Even the *C* mutant of *Calliphora*, which lacks both pterines and ommochromes, contains electron dense carrier granules in the eye-pigment cells (Langer, 1967). Caspari and Richards (1948) and Hanser (1948) suggested, on the basis of a pyronin stain reaction, that the pigment granules consist of ribonucleoprotein. Granules concentrated through differential centrifugation (pigmented in the wild type, colourless in mutant *W* of *Drosophila melanogaster*) proved to consist of an RNA/ protein complex and to contain a certain amount of phospholipid, very similar to the composition of mitochondria (Ziegler and Jaenicke, 1959).

Electron microscopy of *Drosophila* and *Ephestia* clearly shows that the ommochromes and the pterines are located on separate granules (Maier, 1965; Shoup, 1966). The ommochrome granules are homogeneously electron dense and stain strongly with osmium, the pterine granules show a fibrillar or lamellated fine structure, especially visible in the developmental stages when the pterines are being deposited (Maier, 1965; Nolte, 1961; Shoup, 1966; Ziegler, 1960b). Nolte (1961), however, seems to think that in the central and proximal parts of the secondary pigment cells both pterines and ommochromes are found on the same granules. The most recent work shows that the pigment granules are surrounded by a membrane (Shoup, 1966; Langer, 1967). Shoup (1966) described a close connection between the granule synthesis and the Golgi body although Fuge (1966) found no evidence for such a connection.

The establishment of the presence of two components in the pigment granules, i.e. pigment and carrier protein, should not lead to an independent consideration of these components. They are, in fact, fully interdependent. This is shown at the time of granule synthesis, and in situations as found in pigment-free mutants, for instance, in mutant *C* of *Calliphora erythrocephala* one finds only granules of lower electron density and irregular size, which are the pigment free granule "relics" of Langer (1967). During adult development there is no stage where unpigmented granules can be found, as granule synthesis always involves the simultaneous formation of the protein granule and the deposition of the pigment (Maier, 1965; Fuge, 1966). During granule formation, a gradual change in electron opacity is observed (Shoup, 1966) which is considered indicative of alternation in chemical states during development, suggesting pigment synthesis on the surface of the growing granule. Instead of the normal lamellated structure of pterine granules, as found in wild type *Drosophila*, the granules in the pterine lacking mutants *W* or *cn/bw* are irregular in form, resembling autophagic vacuoles of degenerating tissue (Shoup, 1966).

Muth (1967), using *in vivo* application of 3-hydroxykynurenine in the *a* mutant of *Ephestia kühniella,* showed a distinct precursor concentration effect on ommochrome synthesis based, in part at least, on an increased ommochrome synthetic potential in the carrier granule. Synthetic stimulation not only of an enzyme-complex substrate, but also of a structural unit such as a carrier granule, would be a most interesting developmental phenomenon. Similar experiments concerning pterine synthesis have not yet been undertaken because of the very incomplete understanding of the biochemical pathways leading to eye pterine synthesis.

A close connection is also found between sepiapterin and its carrier protein in the epidermal pigment granules of *Bombyx* larvae. Tsujita and Sakurai (1963) have shown that, besides the large amount of epidermal sepiapterin in the mutant *lemon,* there is also a change to a different carrier protein, different both in electrophoretic properties and in peptide composition. This carrier protein of *lemon* has a raised sepiapterin binding potential. A third protein, which is intermediate in pterine binding potential, is found in the mutant *dilute lemon.* When the epidermal pigment granules are destroyed *in vivo*, through the application of 2,4,6-triamino-1,3,5-triazin, the sepiapterin in the cells is temporarily replaced by a blue fluorescing pterine, after which all pterines seem to disappear from the epidermis.

The close connection between pterines and carrier granules is further evident in the optic behaviour of sepiapterin in the *se/se; v/v* mutant of *Drosophila.* In this mutant the ommochromes are absent, and sepiapterin is accumulated. At the time of emergence the pigment granules are bright yellow, but within 24 hours they change to dark brown (Danneel, 1955). This colour change appears to be the result of a complex-formation between the pterine and a tryptophane containing protein of the carrier granule. At the time of this complex-formation, a bathochromic shift can be observed *in vitro* (Fujimori, 1959).

Some doubt has arisen concerning the universality of the protein granule as the carrier of eye pterines. In *Calliphora erythrocephala* mutant *W* the ommochromes are lacking (Tate, 1947; Hanser, 1959), yet both sepiapterin and tetrahydrobiopterin are present (Ziegler, 1961a; Patat, 1965). The diffuse yellow colour of the retinula cells of this species (Hanser, 1959) and Langer's (1967) microspectrophotometric observations on pigment cell granules which show no sepiapterin absorption maxima, give the impression that in *Calliphora* the pterines are not bound to the carrier granules. There is, however, no indication that the diffuse yellow colour of the retinula cells is based on sepiapterin, while it is distinctly possible that the absence of sepiapterin absorption peaks in the carrier granules is the result of the above mentioned bathochromic shift of pterine–protein complexes.

C. IN OTHER TISSUES

Little is known about the localization of colourless pterines that are found in traces in all insects and in all tissues. Hanser and Rembold (1968) traced ^{14}C-biopterin and ^{14}C-neopterin injected into *Apis* larvae, and found these substances accumulated in several metabolically active tissues, and in tissues at the time of their formation during metamorphosis. The ^{14}C activity was often found associated with the nuclei of such tissues. In the *membrana capsularis* and the septa of the testes of *Ephestia,* the ommochromes are attached to typical carrier granules (Caspari, 1955); whether the gonad pterines are on granules is not known. Nevertheless, in all cases, pterines do not fluoresce *in vivo*. Only after an artificial raising of the pH to above normal physiological levels do pterines fluoresce in the tissues. This indicates the presence of a pterine-protein complex even if microscopically recognizable granules are absent. Furthermore, at the time of deposition of pterines in the wings of the pharate imago of *Pieris,* large quantities of leucopterin and isoxanthopterin are carried in the haemolymph, while in pure form at physiological pH levels these substances are insoluble (Harmsen, 1966c).

VI. METABOLISM OF INSECT PTERINES

The most extensively studied metabolic pathway reaction involving pterines is the hydroxylation of C-7. This reaction was first described for the biotransformation of 2-amino-4-hydroxypteridine into iso-xanthopterin and of xanthopterin into leucopterin, through the activity of xanthine oxidase (Wieland and Liebig, 1944; Krebs and Norris, 1949). A convenient assay for xanthine oxidase activity level in *Drosophila melanogaster* based on fluorometric measurement of the isoxanthopterin concentration after incubation with 2-amino-4-hydroxypteridine was reported by Glassman (1962a). In *Drosophila* xanthine oxidase activity is high in the larval fat body and in the adult haemolymph, but it appears to be absent in both larval and adult testes (Ursprung and Hadorn, 1961; Munz, 1964). The accumulation of isoxanthopterin in the gonads, as in the eyes, can be best explained on the basis of absorption from the haemolymph rather

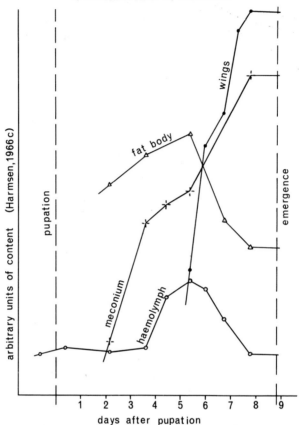

Fig. 3. Internal distribution of leucopterin in the pupa and pharate adult of *Pieris brassicae* (Harmsen, 1966c).

than *in situ* synthesis. It is of interest that the isoxanthopterin concentration of the gonads corresponds with a higher level in the ♂ than in the ♀ eyes. The increased amount of isoxanthopterin in ♂ eyes is unique in that the larger ♀ eyes contain higher quantities of all other pterines. Transplanted eye-imaginal buds and testes are host-specific (Hadorn and Ziegler, 1958; Hadorn *et al.*, 1958), which further suggests that isoxanthopterin is carried to these organs as such. Since ♂♂ and ♀♀ of *Drosophila* have the same levels of xanthine oxidase activity, the higher amounts of isoxanthopterin in the ♂ probably depend on a higher storage capacity (Munz, 1964).

In *Bombyx mori*, xanthine oxidase activity is found throughout the life cycle (Nawa *et al.*, 1957). In *Drosophila* the enzyme activity has a peak in 3 day old flies, and reaches a minimum in the early pupa (Munz, 1964).

Insect xanthine oxidase, which is not easily oxidized and has the same affinity for 2-amino-4-hydroxypteridine and xanthine, appears to be a different species of protein from the mammalian enzyme which is easily oxidized and has a preferential affinity for purines (Nawa *et al.*, 1957).

The mutants *rosy, maroon-like* and *bronzy* of *Drosophila melanogaster* have proved to be most valuable tools in the study of xanthine oxidase and its role in insect metabolism. These mutants completely lack xanthine oxidase activity, and consequently lack both isoxanthopterin and uric acid (Hadorn and Schwinck, 1956; Forrest *et al.*, 1956), showing instead an accumulation of hypoxanthine and 2-amino-4-hydroxypteridine, as well as a number of C-6 substituted eye pterines not oxidized in C-7. A further characteristic of these mutants is the much reduced content of drosopterins (see also p. 181). Transplants of xanthine oxidase active tissues, such as fat body, from wildtype *Drosophila* or from suitable mutants such as *white, brown* or *cardinal* into the larvae of one of the enzyme deficient mutants will immediately stimulate the synthesis of isoxanthopterin in the recipient (Hadorn and Schwinck, 1956; Hadorn and Graf, 1958). It is difficult to explain how such transplants also result in an increase in drosopterin synthesis. Similarly, xanthine oxidase inhibition with 4-hydroxypyrazolo (3,4-d) pyrimidine in wildtype *Drosophila* will phenocopy the enzyme deficient mutants such as *ry* and *ma-1* in respect to decreased drosopterin synthesis (Keller and Glassman, 1965; Boni *et al.*, 1967). It appears that the enzyme activity deficient mutants lack an essential co-factor rather than the enzyme itself. Fatbody transplants supply the recipient with the necessary co-factor. The analysis of the enzyme, its properties and activity in various mutants, has been extensively studied (Glassman and Mitchell, 1959; Glassman, 1962b, 1965, 1966).

Forrest *et al.* (1961) extracted an enzyme from *D. melanogaster* which is similar to xanthine oxidase, but it also catalyses the reaction 4-hydroxy-pteridine → 2,4-dihydroxypteridine (lumazine). It appears that this enzyme is part of the pterine to lumazine pathway (see below), rather than being involved in a hypoxanthine → pterine pathway, as speculated by Forrest and co-workers (1961). In general, xanthine oxidase affects positions 2,4 and 7 of the pterine ring. Previously it was thought that the 6 position of the pterine ring could not be affected by xanthine oxidase (Bergman and Kwietny, 1959). Recently, however, Rembold and Gutensohn (1968) have found that a xanthine oxidase, isolated from rat liver or milk will catalyse the hydroxylation of C-6 of tetrahydrolumazine, tetrahydrofolic acid and tetrahydrobiopterin. The immediate substrates for the enzyme activity seem to be the partly oxidized 7,8-dihydro substances, where the C-5 to C-6 double bond is attacked. This action of xanthine oxidase is also considered to be the basis of the synthesis of xanthopterin from 2-amino-4-hydroxypteridine via the respective 7,8-dihydro forms (Rembold and Gutensohn, 1968), see also p. 182.

A pterine de-aminase, which catalyses the deamination of 2-amino-4-hydroxy pteridine, was extracted from *Alcaligenes faecalis* (Levenberg and Hayaishi, 1959). This enzyme may have an important role throughout the animal kingdom in that it appears to affect all the C-6 substituted pterines (biopterin and related substances) and thus plays a role in the pterine to lumazine pathway (Plaut, 1964). Aruga *et al.* (1954) found an enzyme in the epidermis and fat body of *Bombyx* which transformed "xanthopterin B_1" into "xanthopterin B_2". Recently, Tsusue (1967) has shown that this enzyme is a de-aminase which acts on sepiapterin (xanthopterin B_1). The end product of this deamination (xanthopterin B_2) was identified as 7,8-dihydro-6-lactoyl-lumazine (Goto *et al.*, 1966).*

Deamination is also the first step in the pterine to purine pathway which is followed by hydroxylation of the pyrazine ring and eventual "isomerization" into xanthine-8-carbonic acid. The details of this reaction have only been described for bacteria (McNutt, 1963, 1964). An investigation, especially on non-pteridine accumulating insects where "ring contraction" rather than "ring expansion" of pterine synthesis (see Section X, p. 177) can be expected to take place, could be very instructive.

Of special biological importance are the reactions involved in hydrogenation and dehydrogenation of the C-6 substituted pterines. Sepiapterin from *Drosophila* eyes can be reduced both chemically (Ziegler and Jaenicke, 1959; Ziegler, 1960a) and enzymatically (Taira, 1961a, b; Matsubara *et al.*, 1963) with $NADPH_2$ serving as electron donor (Fig. 4). The enzyme used for this reaction was originally dihydrofolic acid reductase prepared from chicken liver (Osborne and Huennekens, 1958). A similar enzyme has since then been extracted from *Drosophila* prepupae (Taira, 1961b) and *Bombyx* (Tsujita, 1961; Matsubara *et al.*, 1963). This sepiapterin reductase appeared to be different from either folic acid reductase or dihydrofolic acid reductase (Matsubara *et al.*, 1963); indeed, both sepiapterin reductase and folic acid reductase are present in *Bombyx* larvae, but their activity levels fluctuate independently from one another. Folic acid reductase is not present in the pupa, while sepiapterin reductase

* Recently, Rembold and Simmersbach (1969) have isolated a pterine deaminase from rat liver, which is different from guanase. This enzyme deaminates a variety of pterines, producing the corresponding lumazines. What appears to be a similar enzyme was isolated from bee larvae.

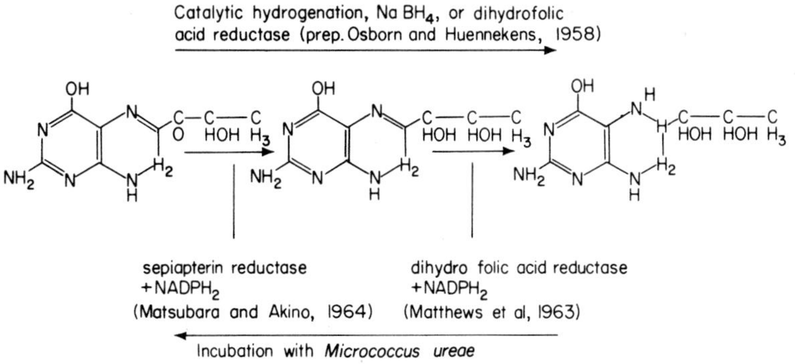

Catalytic hydrogenation, Na BH$_4$, or dihydrofolic
acid reductase (prep. Osborn and Huennekens, 1958)

sepiapterin reductase   dihydro folic acid reductase
+NADPH$_2$ +NADPH$_2$
(Matsubara and Akino, 1964)   (Matthews et al, 1963)

Incubation with *Micrococcus ureae*

Fig. 4. Chemical and enzymatic transformations of sepiapterin, dihydrobiopterin and tetrahydrobiopterin.

reaches a peak between 12 and 15 days after pupation. A specific sepiapterin reductase was also found in rat liver (Matsubara and Akino, 1964). This enzyme is an NADPH dependent reductase which catalyses the hydrogenation of sepiapterin into dihydrobiopterin (through a reduction of the keto group in the side chain), (Matsubara *et al.,* 1966). Dihydrobiopterin can be further reduced to tetrahydrobiopterin under the influence of dihydro folic acid reductase. It must be stressed, however, that until now no biopterin biohydrogenation has been found. Insects appear to be incapable of re-entering biopterin into the active co-factor pool.

The accumulation of sepiapterin in the epidermis of the *lemon* mutant of *Bombyx* is based on the absence of sepiapterin reductase (Tsujita, 1961) combined with the presence of a special carrier granule protein (see p. 164). The folic acid reductase is fully present in *lemon;* only the sepiapterin reductase is absent (Matsubara *et al.,* 1963). The appearance of the egg and young larva depends greatly on the genotype of the mother. This maternal effect is based on the enzyme activity of the ovary. In the mutant *Hn* of *Drosophila,* which also accumulates sepiapterin, a specific inhibitor of sepiapterin reductase seems to be present (Taira, 1961c).

When pure sepiapterin is reduced with liver sepiapterin reductase, and subsequently treated with dilute sulphuric acid in the presence of atmospheric oxygen, a certain amount of isosepiapterin appears (Katoh and Akino, 1966). It is not known if and how an analogous reaction could take place *in vivo.* It is even possible that all isosepiapterin isolated and identified (always in very small quantities) is an artifact, resulting from oxidation of tetrahydrobiopterin during extraction and chromatography.

Oxidation of synthetic 2-amino-4-hydroxy-5,6,7,8-tetrahydro-pteridine through exposure to atmospheric oxygen produces red condensation products (Viscontini and Weilenmann, 1959). It is possible that the drosopterins are synthesized similarly through an enzymatic dehydrogenation of di- or tetrahydropterines. Such a reaction has not been described for insects, but incubation of *Micrococcus ureae* with sepiapterin will result in the synthesis of drosopterins (Nathan and Ziegler, 1961).

VII. HYDROGENATED PTERINES AS CO-FACTORS

In a series of publications, Kaufman (1958, 1959, 1963, cf. 1967) has reported on his work on the enzymic hydroxylation of phenyl-alanine to tyrosine. Synthetic dimethyl-tetrahydropterin was found to act as a necessary co-factor with a rat liver enzyme preparation. A complete enzyme system needing the pterine co-factor in catalytic quantities only was composed of both rat and sheep liver extracts, $NADPH_2$ and a tetrahydropterin (Fig. 5). Dihydrobiopterin was

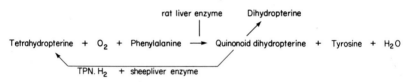

Fig. 5. The role of hydrogenated pterines in enzymic hydroxylation of phenylalanine (after Kaufman, 1967).

isolated as the naturally occurring hydroxylation co-factor from rat liver; *in vivo* it is probably present in its tetrahydro form. This isolated co-factor first has to be reduced to the tetrahydro-level by dihydrofolate reductase.

Sepiapterin from *Drosophila* eyes will also act as co-factor in a complete enzyme system: the K_m value here being $3-4 \times 10^{-6}$ M. Isosepiapterin in low concentrations (0.001 μmol/ml) has approximately 1/3 the activity of the natural co-factor; in higher concentrations it inhibits the system (Kaufman, 1962). The previously discussed work of Matsubara and co-workers (1966) shows that sepiapterin itself does not act as a co-factor; sepiapterin reductase (with $NADPH_2$) produces dihydrobiopterin, which in turn is reduced under the influence of dihydrofolic acid reductase. It is, therefore, the resulting tetrahydrobiopterin which is the real co-factor in phenylalanine hydroxylation. In consequence, in the incomplete system (without sheep liver extract), only tetrahydrobiopterin will act as co-factor and only in stoicheiometric amounts (Ziegler, 1963b).

The complete system has been shown to be present in extracts of whole larvae of *Celerio euphorbiae, Bombyx mori, Ephestia kühniella* and *Dixippus morosus* (Belzecka *et al.*, 1964); these authors failed to find any activity in the pupae of a variety of holometabolous insects. Guroff and Strenkoski (1966) have shown that some of the phosphorylated, hydrogenated intermediates of the purine to pterine transformation pathway (Fig. 6) are active as phenylalanine hydroxy-

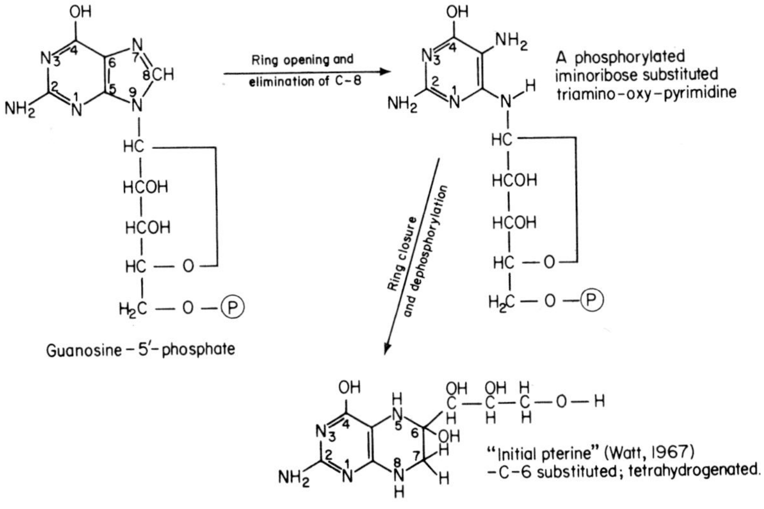

Fig. 6. Biosynthesis of pterines from the purine ring (guanosine-5[1]-phosphate).

lation co-factors. It must be pointed out here, that several hydrogenated pterines have been found to be active, but it is not known which pterine is the natural co-factor *in vivo*.

Also the hydroxylation of tyrosine to dopa is dependent on an hydrogenated pteridine co-factor (Brenneman and Kaufman, 1964; Ellenbogen *et al.*, 1965). Ikeda and co-workers (1967) have shown in an adrenal enzyme system that both steps (phenylalanine→tyrosine →dopa) are catalysed by the same hydroxylating system.

Since dopa leads towards the synthesis of catecholamines, it is of especial interest to note that *Pieris* larvae contain relatively high concentrations of norepinephrin (Euler, 1961) and that Furnaux and McFarlane (1965) have isolated dopa, dopamine and n-acetyl-dopamine from the eggs of *Acheta domesticus*. The function of catecholamines in insects is still only poorly understood, although it has been established that these substances are most probably not involved in synaptic transmission. The role of hydrogenated pterines in the insect's internal co-ordinating systems is probably an essential one, but not a part of the functioning of the nervous system itself (Colhoun, 1963; Davey, 1964).

Recently it has been established that tetrahydrobiopterin functions as a co-factor in the hydroxylation of tryptophane in mammalian brain tissue (Gal *et al.*, 1966; Lovenberg *et al.*, 1967). This hydroxylation is the first step in the synthesis of such substances as 5-hydroxy tryptamine and serotonin. Consequently, the tetrahydrobiopterin is closely connected with the brain. The pharmacological effects of serotonin in the insect suggest that this substance or related substances are natural internal transmitter units: this again stresses the importance of the pterines in the insect's co-ordinating system. The above considerations may be related to the observed accumulation of exogenously supplied ^{14}C-biopterin and neopterin in the ganglion cells of *Apis* (Hanser and Rembold, 1968).

Whether hydrogenated pterines are involved in the hydroxylation of kynurenine is not certain (Ghosh and Forrest, 1967a). Kynurenine hydroxylase has been studied in some detail in mammals (Weber and Wiss, 1966), but the insect enzyme has only recently been isolated by Linzen and Hertel (1967) from *Calliphora* and *Bombyx* larvae and by Ghosh and Forrest (1967a) from *Drosophila*. The activity level of this enzyme varies greatly in insects and is dependent upon as yet unknown factors.

There is some indication that in Protozoa the de-saturation of fatty acids and the ω-oxidation of long-chain fatty acids depends on pteridines. This is probably related to the initial hydroxylation in both these cases (Dewey and Kidder, 1966). There also seems to be a pterine involvement in pyrimidine synthesis in *Crithidia fasciculata* (Kidder and Dewey, 1963). The above mentioned reactions with pterine co-factor involvement have not yet been studied in insects.

VIII. RELATION TO OTHER PIGMENTS

A. MELANIN AND SCLEROTIN (= PROTEIN TANNED BY QUINONES)

Dark pigmentation in insects, especially in the cuticle does not necessarily depend on the presence of true melanin. In many cases "melanin" pigmentation has proven to be based on the presence of ommochromes. In the exocuticle true melanin seems to occur, together with sclerotin (cf. Hackman, 1964). Both are derived from tyrosine and dopa. In poikilothermal vertebrates a close morphological and ontogenetic relationship between melanin production and the occurrence of tetrahydrobiopterin is obvious (Ziegler, 1963b; Kokolis and Ziegler, 1968). This fact is understandable considering the role of hydrogenated pteridines in the initial steps of melanin synthesis, i.e. the hydroxylation of phenylalanine and tyrosine. Even though this particular role of tetrahydropterines in insects is suggested by the findings of Belzecka et al. (1964), definite information concerning these reactions in insects is completely lacking.

Also polyphenols taking part in colouring and hardening of insect cuticle arise from dihydroxy-phenols and thus originally from phenylalanine (cf. Hackman, 1964). Investigations on participation of pterine co-factors on this pathway are completely lacking.

The findings of Hanser and Rembold (1968) showing that ^{14}C-biopterin and neopterin, fed during the larval stages of Apis, result in the accumulation of ^{14}C activity in the exocuticle of 3 day old pupae, are hard to explain. Exogenous biopterin is previously reported not to be metabolized (Rembold and Hanser, 1960a); if this is so, it can neither be involved as co-factor in the first hydroxylation steps, nor in polyphenol synthesis. Moreover these processes seem to take place inside the insect's body and complete phenols diffuse outward through the cuticle to the epicuticle where they undergo final oxidation (cf. Hackman, 1964). The possibility that pterines are deposited in the exocuticle (as in the epidermal cells) because it is a preferred site for deposition of metabolic end-products cannot be ruled out.

B. OMMOCHROMES

Our knowledge of the biological relationship of pterines and ommochromes is rather the reverse of the situation with melanin. In the case of the ommochromes, pterine involvement in the hydroxylation of kynurenine is still highly speculative (Ghosh and Forrest, 1967a, b), while much evidence exists, both histological and genetical, concerning a close relationship (cf. Ziegler, 1961b). Particularly indicative of this relationship is the mutant a of Ephestia kühniella. In this mutant, kynurenine synthesis (and consequently also ommochrome synthesis) is blocked and, at the same time, the pterine content of this mutant is highly abnormal (Hadorn and

Kühn, 1953; Kühn and Egelhaaf, 1955). Especially noticeable is the accumulation of dihydroekapterin (Viscontini and Stierlin, 1962a, b). Both parabiotic fusion with an a^+ individual and injection of kynurenine result in normal ommochrome synthesis as well as in normalization of the pterine content of the a mutant (Kühn and Egelhaaf, 1955). The mutants of the W series of *Drosophila melanogaster* and *chalky* of *Calliphora erythrocephala* are other examples of mutants lacking both ommochromes and pigment pterines in the eyes. The connection of these two phenomena is not understood. Nothing concrete is known about the metabolic aspects of the pterine kynurenine inter-relationship; Ghosh and Forrest (1967a) have suggested that the primary absence of a suitable, hydrogenated pterine secondarily reduces ommochrome synthesis. Ghosh and Forrest (1967b) further speculate that in W of *Drosophila* the tryptophane pyrrolase of the ommochrome biosynthesis pathway is inhibited by the exceptionally high concentration of simple pterines.

In many cases, as in mutant *Wa* of *Ephestia*, the absence of suitable carrier granules is the possible cause of the absence of both ommochromes and pterines (Hanser, 1948). These problems have been discussed in detail by Ziegler (1961b).

C. FLAVINES

In both poikilothermal vertebrates and insects, pterines and riboflavine often occur together; for instance in eggs, eyes and malpighian tubules. The relationship between pterines and riboflavine is particularly well developed in *Formica*, where formicapterin and riboflavine form a chemical complex (Schmidt and Viscontini, 1964). In other cases, however, as in *Melanoplus* (Burgess, 1949), the riboflavine in the diapausing egg is slowly replaced by pterines during development.

The metabolic relationship between riboflavine and pterine is reasonably well understood. The deamination of pterines to produce lumazines has been described above (see p. 168). Lumazine can also be synthesized enzymatically from purines (McNutt, 1956). The further synthesis of riboflavine from lumazine by riboflavine synthetase has been described by Plaut (1964):

2 molecules of 6,7-dimethyl-8-ribityl lumazine combine to form 1 molecule each of riboflavine and "x".
Substance "x" has been identified by Wacker *et al.* (1964) as
4-(1^1-D-ribitylamino)-5-amino-2,6-dihydroxypyrimidine.
Various lumazine derivatives have been isolated from insects (Table III).

D. PURINES

Many tissues of larvae and adult insects, as well as extracts of whole eggs, contain both pterines and purines (Hudson *et al.,* 1959; Kürsteiner, 1961; Auf der Maur, 1961; Harmsen, 1966a, c). However, purines in insects are not usually found as crystalline or granular deposits in the integument, as they are in poikilothermal vertebrates. An exception are the Pieridae, in which simple pterines, as well as uric acid, hypoxanthine (Tartter, 1940; Harmsen, 1966b), xanthine and isoguanine (Purrmann, 1940), are found as deposits in the integument.

The close biosynthetic relationship between purines and pterines (see p. 178) does make their common occurrence far from surprising (see p. 189).

IX. DEVELOPMENTAL PHYSIOLOGY

Pterine accumulations are already found in many insect eggs (Table III). It is possible that these pterines are not the product of metabolic activity in the egg cell, but rather originate from the surrounding parental tissues. This, for instance, is the case with predetermination of development of ommochromes, which was shown with $a^+ \times a$ crosses in *Ephestia* (cf. Ziegler, 1961b). A reduction in the riboflavin content and a simultaneous increase in pterine content in post-diapause eggs of *Melanoplus* (Burgess, 1949) and similarly, a reduction in xanthopterin (probably dihydroxanthopterin) simultaneously with an increase in erythropterin in *Oncopeltus* eggs (Forrest *et al.,* 1966), shows that during further embryological development pterine metabolism becomes active.

In all insect larvae, at least small amounts of fluorescing pterines are present (Table IV). In *Oncopeltus,* for example, both iso-xanthopterin and xanthopterin increase throughout larval development reaching a maximum of approximately 10 μg per individual (Hudson *et al.,* 1959). Obviously erythropterin is also present in *Oncopeltus,* but the extraction technique of Hudson and co-workers (1959) did not allow for an estimation of this substance.

In *Pieris brassicae* the amount of pterines during the last larval instar increases from about 70 μg to about 130 μg per animal (Harmsen, 1966c); most of this pterine deposit is in the integument. As in the silkworm, the larva of *Pieris* has deposits of sepiapterin in its integument and this sepiapterin disappears completely on pupation (Harmsen, 1963). Honeybee larvae reach a maximum concentration (50 μg/g fresh wt.) of biopterin two days after eclosion, but it drops considerably in later instars (Hanser and Rembold, 1960). Prior to pupation, a certain amount of pterine is excreted with the last larval excreta both in *Apis* (Hanser and Rembold, 1960) and in *Pieris* (Harmsen, 1963, 1966c).

In all pterine accumulating species, whether in the integument or in the eyes, the bulk of the imaginal pterines are formed in the latter half of the pupal stage. Harmsen (1966c), Descimon (1966) and Watt and Bowden (1966) followed the abundance of pterines in various parts of the bodies of respectively *Pieris brassicae, Colias croceus* and *Colias eurytheme.* Figure 3 supports the opinion that pterine synthesis takes place in the fatbody, and that the simple pterines are carried via the haemolymph (which temporarily shows a peak in pterine concentration) to the wings and body scales. A pH value of 6–6.5 as measured for pharate imaginal wings would indeed be a perfect environment for the precipitation of crystalline pterines (Becker, 1937b). Becker studied the histology of *Vespa* and *Pieris* at the time of pterine deposition, and concluded that the pterines are brought to the deposition sites, not synthesized *in situ.* The deposition occurs simultaneously in all parts well after cuticle formation, including melanin synthesis, is completed. At the time of maximum pterine deposition, the epidermal cells and their nuclei are already much reduced and apparently inactive: in the wings, the upper and lower integumental surfaces have already come together, reducing the enclosed space to a haemolymph filled system of capillaries.

Contrary to the completely passive role of the wing epidermis in pterine deposition stands the work of Watt (1967), who found that isolated *Colias* wings were capable of guanosine→pterine transformation (see p. 179). Xanthopterin and erythropterin in particular appear to be newly synthesized in the wings, while leucopterin and perhaps also isoxanthopterin seem to be synthesized mainly in the fatbody, as was shown by the relative rates of ^{14}C incorporation (see p. 183).

The eye pterines of Diptera appear late in the development of the pharate imago (70 hours in *Drosophila*), after the ommochromes. Also in *Drosophila*, Zeutzschel (1958) found the first red pigment granules in the fourth day after pupation. After that, pterine deposition takes place rapidly until adult emergence, but the total pterine content of the eyes keeps increasing until 2–3 days after emergence (Hadorn and Ziegler, 1958). In *Ceratitis capitata* (Ziegler and Feron, 1965) and in *Calliphora erythrocephala* (Patat, 1965) the main period of eye pterine accumulation is after emergence of the imago. Isoxanthopterin is an exception, as it appears in *Drosophila* eyes already 20 hours after pupation and, after a brief period of increase, it remains at a constant level from 50 hours onward. It is also exceptional in that it is not bound to the later appearing pigment granules (see p. 162). A further point of interest is that isoxanthopterin is present in twice the concentration in the smaller ♂ eyes than in ♀ eyes (Hadorn and Ziegler, 1958). From reciprocal transplant experiments it appears that the isoxanthopterin level of the abdominally inserted eyes is completely controlled by the host (see also p. 166).

In *Ephestia,* the number and intensity of fluorescing spots (mainly pterine) on whole pupal extract chromatograms increase during development; certain particular substances, however, only occur temporarily (Reisener-Glasewald, 1956). The chemical nature of most of these substances has not yet been established. It is becoming more and more of a profitable prospect to attempt the integration of our increasing knowledge of chemical structure and biosynthesis of pterines with the understanding of specific sequences of occurrence and the cytogenesis of carrier granules as described by Hanser (1948) and Reisener-Glasewald (1956).

X. BIOSYNTHESIS

A. BIOSYNTHESIS OF THE PTERIDINE RING

Although insects appear to be dependent on dietary folic acid (see Ziegler, 1965) they are capable of the biosynthesis of unconjugated pteridines. In certain insect larvae, and even some non-pterine accumulating adults, it is possible to account for all pterines as originating with dietary folic acid. Harmsen (1966c) estimated that a total of 150 μg of folic acid may be taken in with the food of the larva of *Pieris brassicae;* it is possible that the measured 180 μg total pterine content of fully grown larvae can thus be accounted for. However, Jones (1968) has reared *Pieris* larvae on a chemically controlled diet with a folic acid content much lower than in cabbage leaves, and the pterine level in mature larvae reared on this diet was normal. The subsequent increase in pterine content by 500–600 μg during the pupal stage can only be accounted for by biosynthesis from a non-pterine precursor. Feeding experiments with ^{14}C-labelled biopterin have shown that honey bee larvae are incapable of metabolizing dietary biopterin: the entire labelled addition was excreted (Rembold and Hanser, 1960a).

In a more recent work (Hanser and Rembold, 1968) it has been reported that a small amount of dietary ^{14}C-biopterin may become incorporated into the hydrogenated pterine pool, as was shown autoradiographically. The amount which is retained is quite small as is supported by the findings of quantitative feeding experiments: the biopterin content of bee larvae does not differ significantly whether the larvae received 40 μg, or 200 μg biopterin/g dry weight food. Therefore, it appears that insects with a low pterine content (such as the honey bee) also synthesize their own requirement rather than depend on uptake.

It has been shown that the purine ring of guanine can be a precursor for pterine biosynthesis. The guanine→pterine biosynthetic pathway appears to be generally distributed in the animal kingdom (see Ziegler, 1965). It was first described for Pieridae, using ^{14}C-labelled precursors (Weygand et al., 1961; Simon et al., 1963; Schmidt, 1965; Watt, 1967). The pathway is represented in Fig. 6. The purine ring is opened and C-8 is eliminated. This elimination has been proven unequivocally, and appears to take the form of formate (Levenberg and Kaczmarek, 1966; Burg and Brown, 1968), which can either be excreted (CO_2) or re-enter purine synthesis (Watt, 1967). The resulting intermediate is a hydrogenated and phosphorylated pyrimidine; on ring closure it is transformed into a non-fluorescing "initial pteridine" (Watt, 1967) similar to tetrahydrobiopterin. Rembold and Buschmann (1963b) consider neopterin (with a D-erythro side chain) one of the first stable products of the purine→ pterine transformation.

Two substances, which may well represent pterine precursors, have been isolated from *Pieris* pupae (Schmidt, 1965). These substances completely disappear after the deposition of wing pterines (Ziegler, unpublished). Similar, non-fluorescing, yet *Crithidia*-active substances have been isolated from *Drosophila* larvae, and are particularly abundant in the haemolymph of freshly emerged adults at a time when eye pterine synthesis is at its peak (Ziegler, unpublished). Such non-fluorescing *Crithidia*-active substances have also been found in larvae of *Apis* (Rembold and Hanser, 1960a). At this stage it seems that a reliable quantitative analysis of these pterine precursors could reveal much concerning the site and timing of pterine synthesis, and may even add to our understanding of the role of pterines in the entire metabolic process (see p. 186).

Becker's (1937) results from rearing *Pieris* pupae in pure oxygen and thus producing pterine free adults have been interpreted as indicating the presence of a labile, hydrogenated precursor sometime in the pupa. Recent rearing experiments at different oxygen levels of pupae of *Mylothris chloris,* however, do not substantiate Becker's results (Harmsen, 1969a).

Although it appears probable that all, or most, pterine biosynthesis involves a purine to pterine transformation, it is not certain whether only completely formed purines or also common purine precursors are involved. Weygand and co-workers (1961) and also Simon and co-workers (1963) consider known, completely formed purines the most likely pterine precursors. They base their opinion on specific radioactivity and on incorporation rates of glycine (2-^{14}C) and the purine precursor amino-imidazole-carboxamide (4-^{14}C). They also point out that the total content of adenine and guanine drops from 720 μg in the prepupa to 250 μg at emergence; during the same period, the leucopterin content increases by 500–750 μg. For this relationship to have any significance at all, one would have to assume that no other purines are involved in pterine formation, and that adenine and guanine are only precursors for pterines; furthermore, one would have to exclude any further adenine and guanine synthesis. In fact, Harmsen (1963) measured an increase of 1200 μg xanthine and 500 μg uric acid during metamorphosis, showing that in *Pieris* during the pupal stage both pterine and purine content increase considerably. The rather complex situation of purine to pterine transformation taking place simultaneously with purine synthesis and with internal transport of both purines and pterines is obviously not a promising one from which to elucidate synthetic sequences based on relative concentrations.

B. BIOSYNTHESIS OF C-6 SUBSTITUTED PTERINES

Feeding experiments with glucose-6-^{14}C and 2-^{14}C in *Drosophila* larvae have shown that C-2 and C-3 of the side chain of biopterin and the drosopterins originate from C-5 and C-6 of the glucose (Brenner-Holzach and Leuthardt, 1961, 1967). Watt (1967) injected labelled guanosine into 96 hour old pupae of *Colias eurytheme* and incubated isolated pupal wings with the same substance; he subsequently studied isotope incorporation kinetics for sepiapterin from four minutes to several hours after injection. Oxidation of the sepiapterin to 2-amino-4-hydroxypteridine-6-carboxylic acid and determination of the specific radioactivities of the parent compound as well as the oxidation product, showed decisively that the C-6 side chain of sepiapterin is derived from guanosine ribose. Therefore, the side chain is not secondarily attached, but originates at the time of pterine synthesis, and can later be modified, substituted or eliminated. Radioactive pyruvate, malate and lactate, administered in the same way, are not incorporated into the sepiapterin, in contrast to erythropterin (see p. 182). These experiments corroborate analogous work on the biosynthesis of C-6 substituted pterines in microorganisms (Reynolds and Brown, 1964).

Contrary to the above results, McLean *et al.* (1965) have reported the incorporation of labelled threonine into the pterine ring at C-6 in the alga *Anacystis*. These authors postulate a similar biosynthetic pathway for the drosopterins. Watt (1967), however, has pointed out that the very low level of incorporated isotope and the long duration of the experiment may very well have resulted in a general labelling of the *Anacystis* carbon pool.

Sugiura and Goto (1967) fed guanosine-5-phosphate-^{14}C to larvae of *Drosophila melanogaster* and subsequently analysed the pteridines extracted from the whole flies. They also found guanosine-5-phosphate to be an effective precursor of biopterin and of 2-amino-4-hydroxypteridine. This Japanese group (Goto *et al.*, 1964; Okada and Goto, 1965; Goto *et al.*, 1965; Sugiura and Goto, 1967) also fed a variety of ^{14}C labelled pterines, such as neopterin, hydrogenated 2-amino-4-hydroxypteridine and 2-amino-4-hydroxy-6-hydroxymethylpteridine with the food to *Drosophila* larvae. Analysis of chromatograms of whole flies for ^{14}C activity in pterines made these authors conclude tentatively that the original C-6 three carbon side chain is detached and that the C-6 side chain of biopterin and related substances is secondarily attached, and is, as yet, of unknown origin. The long and metabolically highly active period between intake and analysis (the entire metamorphosis) and the inaccuracies of chromatographic identification make any conclusions based on this work highly suspect.

Viscontini and Weilenmann (1959) obtained red condensation products on the oxidation of 2-amino-4-hydroxy-5,6,7,8-tetrahydropteridine. They speculated that this *in-vitro* reaction may be representative of the *in-vivo* synthesis of the red eye pterines. Analysis of paper chromatograms of eye extracts throughout development of various *Drosophila* strains and mutants (cf. Ziegler, 1965) has shown that drosopterin synthesis probably involves a dehydrogenation of tetrahydrobiopterin via sepiapterin: the tetrahydro- and dihydro-compounds appear in sequence prior to the deposition of the red pterines (Hadorn and Ziegler, 1958). Moreover, in those mutants where tetrahydrobiopterin and sepiapterin accumulate (no drosopterins are formed), a gradual increase in ratio sepiapterin/tetrahydrobiopterin indicates a sequential dehydrogenation (Ziegler and Hadorn, 1958; Ziegler and Feron, 1965): pterine precursor→ tetrahydrobiopterin→sepiapterin→drosopterins. However, the occurrence of tetrahydrobiopterin as an excretory product in the meconium (Wessing and Eichelberg, 1968) and the relative abundance of hydrogenated and phosphorylated pterine precursors in the haemolymph, suggest the possibility of parallel rather than sequential synthesis of tetrahydrobiopterin, sepiapterin and the drosopterins:

⟋tetrahydrobiopterin
pterine precursor→sepiapterin
⟍drosopterins

It should be possible to corroborate the postulated dephosphorylation and dehydrogenation of the initial pterine as carried to the eyes, through a study of the dehydrogenating enzymes involved. The relative activity levels and timing of activity of the various dehydrogenases (Ziegler and Jaenicke, 1959) in the eye would probably indicate the nature of the synthetic pathway of the drosopterins. Also, the ommochromes appear to deposit on the granules through an "oxidative condensation" process (Butenandt et al., 1956; Butenandt, 1959).

A close relationship exists between drosopterin synthesis and the extent of phenylalanine hydroxylation in the mutants rosy- and maroon-like of Drosophila melanogaster. Wounding of the pupal case of these mutants (especially when followed with phenylalanine incubation) results in enhanced melanin synthesis; simultaneously, the normally very low drosopterin content increases. Tyrosine and dopa have no such effect (Schwinck, 1965). It is possible that the tetrahydro co-factors of the hydroxylases become irreversibly dehydrogenated, eventually leading to drosopterins. New co-factors may not be formed through re-hydrogenation of quinonoid dihydropterine (Fig. 5) but through completely new pterine synthesis. It should be determined whether the increased drosopterin synthesis in rosy reared at lowered temperature (Hadorn and Schwinck, 1956) is also associated with an increase in phenylalanine hydroxylation.

C. BIOSYNTHESIS OF OTHER PTERINES

The elegant experimental work of Watt (1967) shows that in *Colias eurytheme* the three carbon side chain of the initially synthesized pterine persists, at least in part, as the side chain of sepiapterin and probably of other C-6 substituted pterines. The simple pterines are formed through the irreversible, oxidative elimination of this side chain (Rembold *et al.,* 1969).

Isoxanthopterin, xanthopterin and leucopterin are formed through further oxidation of C-6 and C-7 positions of 2-amino-4-hydroxy-pteridine (Gutensohn, 1968; Rembold and Gutensohn, 1968; Harmsen, 1969) (see below). The synthesis of erythropterin (a C-7 substituted pterine) appears to be based on the addition of a side chain onto xanthopterin (Forrest *et al.,* 1966; Watt, 1967). This side chain can originate from pyruvate, malate, or lactate and at a lower rate from serine (Watt, 1967); Forrest and co-workers (1966) suggested oxalo-acetic acid as the metabolic source of the side chain 3-C unit. The incorporation of a secondary side chain probably involves a more reactive, hydrogenated derivative of xanthopterin. This reactive compound (probably 7,8-dihydroxanthopterin) originates with the breakdown of tetrahydrobiopterin, in the final stage of which xanthinedehydrogenase can act on the C-6 position of 7,8-dihydro-2-amino-4-hydroxypteridine (Gutensohn, 1968). The presence of 7,8-dihydroxanthopterin in the wings of freshly emerged *Colias* (Descimon, 1967) and in the eggs of Oncopeltus (Forrest *et al.,* 1966) at the time of erythropterin synthesis and its subsequent disappearance seems convincing evidence. Forrest and co-workers (1966), however, failed to stimulate erythropterin synthesis in egg extract, incubated with synthetic 7,8-dihydroxanthopterin. In any case, it is quite clear that the C-7 side chain, being a secondary addition, has a very different origin from the C-6 side chain which is a remnant of the original pterine synthesis. This situation corroborates the wide biological and metabolic separation of C-6 and C-7 substituted pterines (Harmsen, 1963, 1964).

Intensive quantitative analysis of the leucopterin and isoxanthopterin content of gut, fatbody, haemolymph and wings of the pupa and pharate imago of *Pieris brassicae* during metamorphosis shows that these pterines are, to a large extent, synthesized in the fatbody and then transported as such, via the haemolymph, to the wings and the meconium (Fig. 3) (Harmsen, 1966c). Watt (1967), however, found the ability of enzymic transformation of guanosine to pterine in the wings of the pharate imago.

However, it is possible that in *Colias* the xanthopterin and erythropterin are freshly synthesized in the wings, while the leucopterin and isoxanthopterin originate in the fatbody. The much higher ^{14}C incorporation in xanthopterin than in leucopterin in *Colias* (Fig. 3 in Watt, 1967) would suggest this. The biosynthetic basis of such a division in simple pterine production must be the extent of hydrogenation of the intermediates of pterine breakdown. All three C-6 and/or C-7 oxidized simple pterines are formed under the influence of xanthine oxidase (Rembold and Gutensohn, 1968). 7,8-Dehydrogenation of di-hydropterines, on the other hand is dependent on the local oxygen levels and may be either non-enzymic or catalyzed by an as yet unknown enzyme. The key substance after chain elimination: 7,8-dihydro-2-amino-4-hydroxypteridine (dihydropterin) can thus be variously oxidized. At low oxygen levels this substance is sufficiently stable to allow xanthine oxidase to act at C-6, resulting in the production of dihydroxanthopterin. At higher oxygen levels, or in the presence of an unknown pterine-7,8-dehydrogenase, dihydropterin is very unstable and will be dehydrogenated to pterin (2-amino-4-hydroxypteridine). In the presence of xanthine oxidase, pterin is subsequently oxidized to isoxanthopterin (Gutensohn, 1968; Harmsen, 1969). If at relatively high oxygen levels a certain amount of dihydropterin is converted into dihydroxanthopterin, this latter substance is oxidized to xanthopterin and eventually to leucopterin. At low oxygen levels, on the other hand, dihydroxanthopterin is stable and leucopterin will not be formed (see Fig. 7). In *Colias* wings, dihydroxanthopterin is probably oxidized to xanthopterin very late in metamorphosis when no more active xanthine dehydrogenase is present. The apparent natural division between *Colias* fatbody and wings has been experimentally copied by Harmsen (1969) in *Mylothris* with varying environmental oxygen levels. It is not understood why fatbody pterine synthesis acts as a system exposed to relatively high oxygen levels, and wing synthesis as if exposed to lower oxygen levels. The presence of a 7,8-dihydropterine dehydrogenase in the fatbody could be the answer.

In *Pieris napi* (and especially in its variety *sulphurea*) sepiapterin is reported to be abundantly present, and appears to be the basis of the yellow pigmentation (Watt and Bowden, 1966). Therefore a comparative study of the pterine synthetic capacity of wings and fatbody in both *Pieris brassicae* and *P. napi* could be very revealing also.

In insects where xanthine oxidase is present, any 2-amino-4-hydroxypteridine formed through oxidative elimination of the

C-6-side chain and dehydrogenation of C-7 and C-8 will be transferred to isoxanthopterin. This appears to be the case in *Drosophila* (Ursprung and Hadorn, 1961), where the isoxanthopterin is formed in the fatbody and is subsequently selectively concentrated in the eyes and some other organs. This is in contrast to the drosopterins which are newly synthesized in the eyes on the carrier granules. All genetic, histological and developmental physiological observations confirm the above described situation. This situation seems to represent a general difference between the biopterin type pterines (C-6 substituted), which are the primary products of pterine synthesis and show only minor modifications and dehydrogenation, and the simple pterines which appear to be end products of pterine metabolism. The C-7 substituted pterines are synthesized through the addition of a secondary side chain after the elimination of the primary C-6 side chain. Synthesis of C-7 substituted pterines probably begins with the hydrogenation of simple pterines or with the elimination of the C-6 side chain from hydrogenated C-6 substituted pterines, in either case delivering reactive intermediates such as dihydropterin. Figure 7 represents the various interrelating pathways of pterine metabolism.

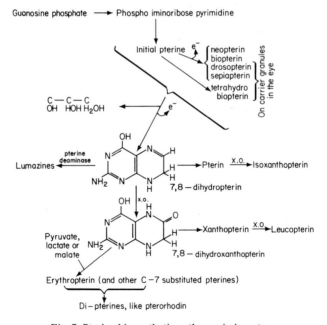

Fig. 7. Pterine biosynthetic pathways in insects.

Of interest for the study of the biosynthesis of erythropterin and pterorhodin are the mutants *a* of *Ephestia kühniella* and *dec* of *Ptychopoda seriata.* Paper chromatograms of eye extracts of *a-Ephestia* and of extracts of dark-reared eyes of *dec-Ptychopoda* show, among others, one blue-green fluorescing spot designated pterine "C_1". On storage of the chromatograms in the dark, this spot turns dark red; the red pigment was identified as pterorhodin (Kühn and Egelhaaf, 1959). All results agree with the concept that an oxidative condensation of a hydrogenated pterine is involved. Such an oxidative condensation can easily take place chemically *in-vitro* (Albert, 1954; Viscontini and Weilenmann, 1959) and is also postulated for the synthesis of drosopterin (see p. 180). Viscontini and Stierlin (1962a, b) have isolated dihydroekapterin from the eyes of *a-Ephestia,* but this substance has not yet been identified with pterine "C_1". *Dec-Ptychopoda,* particularly, is interesting because in this mutant the *in-vivo* synthesis of pterorhodin and erythropterin will take place only when the eye is illuminated in the presence of atmospheric oxygen; the illumination is effective only in wavelengths between 313 and 360 mμ (Hanser, 1948). In such illuminated eyes, pterine "C_1" disappears (Kühn and Egelhaaf, 1959). Similar illumination of animals kept in pure nitrogen has no effect on pterine "C_1". The light dependent pterine synthesis takes place *in-situ* in the eye, one sided illumination results in darkening of the illuminated eye only (Hanser, 1948).

The biosynthesis of folic acid and its derivatives also starts with guanosine or guanosine phosphate. A hydrogenated pterine is produced which is coupled with a p-aminobenzoic acid moity (cf. Jaenicke and Kutzbach, 1963). The reasons why insects are incapable of coupling hydrogenated pterines and p-aminobenzoic acid in the synthesis of the folic acid co-factors, is not known (cf. Ziegler, 1965).

XI. PHYSIOLOGICAL ROLES OF PTERINES IN THE INSECT

A. AS CO-FACTORS AND GROWTH SUBSTANCES

The only presently known definite function of hydrogenated pterines is one of co-factor in hydroxylation reactions (see p. 170). Dehydrogenation and degradation leading to the appearance of fluorescing pterines in the insect nervous system seems to have stimulated speculation that pterines may function as neurohormones

(L'Hélias, 1956) or as sex determinants (L'Hélias, 1961, 1962). Fischer *et al.* (1962) established that the pteridines (isoxanthopterin, xanthopterin and 2-amino-4-hydroxypteridine) extracted from the nervous system of *Periplaneta, Leander* and *Carausius* possess no neurohormonal activity. Their occurrence probably indicates an oxidative breakdown of a biologically active dehydrogenated hydroxylation co-factor, active in catecholamine and serotonin synthesis in the nervous system (see p. 172).

The discovery that dietary biopterin is not metabolized by larvae of *Apis* (Rembold and Hanser, 1960a) destroyed the formerly held speculation (Butenandt and Rembold, 1958) that it was the substance in royal jelly responsible for queen determination (Rembold and Hanser, 1964). The latter opinion was experimentally confirmed later. The comparatively high biopterin concentration of the pharyngeal gland and the accumulation of dietary applied biopterin-2-^{14}C into this organ may reflect a preferential excretory pathway. A similar situation, which may very well be functionally related to the biopterin secretion, is the secretion of 10-hydroxy-Δ^2-decenoic acid by the pharyngeal gland. This substance is synthesized in the mandibular gland and selectively accumulated into, and excreted by, the pharyngeal gland (Rembold and Hanser, 1960b). It is possible that the synthesis of this latter substance by desaturation involves relatively large quantities of a hydrogenated pterine co-factor (Dewey and Kidder, 1966) (see p. 172); the dehydrogenation of the co-factor results in an accumulation of biopterin which is subsequently also excreted by the pharyngeal glands.

B. AS EYE PIGMENTS

Recent experimental work has shown conclusively that light perception in insects also depends on two Vitamin A related visual pigments (Goldsmith, 1958; Briggs, 1961; Goldsmith and Warner, 1964). The study of the role of pterines and ommochromes in the insect eye is of course much aided by the large variety of mutants lacking pterines and/or ommochromes in one way or another. The optomotor response of various mutants of *Drosophila* has been investigated (Kalmus, 1943; Götz, 1964). This work has revealed that a direct relation exists between optomotor response and amount of eye pigment: white eyed mutants (e.g. *W* or *bw:cn*) have no optomotoric reaction and their visual acuity is nil. This can be explained by the increased amount of scattered light in the pigment deficient eyes. The optomotoric responses are also lacking in a white eyed mutant of *Culex molestus*. The relatively larger *Calliphora erythrocephala* mutants *W* (almost ommochrome free and only containing sepiapterin and tetrahydrobiopterin) and *chalky* (completely pterine free) have been investigated not only for visual acuity, but also for spectral sensitivity, retinogram amplitude and respiration rate (Autrum, 1955, 1961; Hoffmann and Langer, 1951; Langer and Hoffman, 1966). The investigation shows no evidence for a function of pterines or ommochromes in the visual process *sensu strictu.* However, these substances play an essential role as screening pigments and in light filtering. The pterines, in particular, function as a light filter in the near UV and blue regions and thus act complementarily to ommochromes which work in the visible region. In the red region neither pigment has a screening effect, which may result in the relative improvement of perceptibility of ineffective red.

C. AS METABOLIC END PRODUCTS

The above discussed metabolic functions of hydrogenated pterines accounts for the general occurrence of these substances and their oxidized end products not only in all insects, but in the entire animal kingdom. In certain groups of insects (and in some poikilothermal vertebrates), the oxidized and/or simple pterines are so abundant that they must represent the end products of a particular section of total nitrogen metabolism, or perform a functional role in their terminal form (Harmsen, 1966a). In this respect they are somewhat similar to the ommochromes (Linzen, 1967).

These simple pterines are usually in a highly oxidized form and are either excreted with the normal excreta (especially in the meconium) or stored in granular or crystalline form in the fatbody or the integument.

The excretion of pteridines is never at a high level, except in insects which ingest large quantities of pterine, such as *Phonoctonus* (Harmsen, 1963). A barrier threshold seems to be active in the Malphighian tubules (see below). A minor, but significant increase in the meconium content occurs in cases where pterine synthesis takes place but where the normal storage sites are disrupted, such as mutant *W* of *Drosophila* with no eye pigment granules (Kürsteiner, 1961), or *Pieris brassicae* variety *caerulea* with a reduced number of wing scales (Harmsen, 1964). This phenomenon has been experimentally induced with pupal wing amputation in *Pieris* (Harmsen, 1966a), which also causes an increase in pterine content of the meconium.

Generally, storage rather than excretion in its traditional meaning occurs when pterines are synthesized in large quantities and not subsequently metabolized. Pieridae have a larger accumulation of pterines than any other family of insects. Any specialized pathway of nitrogen metabolism resulting in the production of relatively large quantities of simple pterines could be expected, therefore, to be best developed in this family. Harmsen (1966a) executed an exhaustive quantitative study of simple pterines in *Pieris brassicae*. In this insect simple pterines constitute 14% of all highly oxidized nitrogen at the time of emergence. Purines account for 80%; the remaining 6% consists of urea, kynurenine and several other minor substances. Of the approximately 200 μg of pterines present at emergence, only 4 μg are excreted with the meconium. This indicates a strong excretion barrier at the time of meconium formation in favour of storage in the fatbody and integument. Experimental removal of one forewing of the pupa of *Pieris* resulted not only in a slight increase of meconial pteridine excretion and fatbody storage, but in a significant drop in the synthesis of simple pterines (Harmsen, 1966a). These results indicate that the synthesis and storage are one continuous process with a controlling feed back. Synthesis is shut off when all storage sites are occupied and the remaining precursor material must be shunted into another (perhaps purine) pathway. The feed back mechanism may be based on a simple concentration gradient. Dehydrogenated, phosphorylated, C-6 substituted, and possibly protein bound pterines are more soluble than simple pterines, in ordinary physiological solutions. Therefore, during deposition in crystalline or granular form, a concentration gradient is maintained. When deposition ceases the concentration gradient of the synthesis sequence is leveled off, and the synthesis reactions are stopped. Furthermore, it is possible that the pterine gradient increases the fluid circulation in the late pupal wings, thereby enhancing the transport of essential substances to the wing tissue.

The development of the increased storage potential and indeed storage preference over excretion, and the increased role of pterines in nitrogen metabolism as displayed in Pieridae, must have evolved simultaneously. Harmsen (1966a) found that the total synthesis of simple pterines in *Vanessa io* is about one tenth of that in *Pieris*. *Vanessa* has virtually no deposition in the wings, and excretes approximately 50% of all pterines with the meconium. In fact, the amount of total pterines excreted by *Vanessa* is higher than in *Pieris*.

Therefore, it seems that *Vanessa* can be considered a "normal" insect which excretes the oxidized end products of the biologically active hydrogenated pterines. In *Pieris,* pterine synthesis is vastly in excess of the normal level necessary for the supply of the physiologically active hydrogenated substances. The very low percentage of *Pieris* pterines which is excreted must reflect an advantage of storage over excretion. Similarly, the development of pterine synthesis as a final step in part of the nitrogen metabolism must reflect an advantage of pterines over purines.

Potentially reversible storage of nitrogenous end products is present in many insects. Shortly after emergence, the fat body of *Pieris* contains 150 μg simple pterines, 1400 μg uric acid, and 1200 μg xanthine (Harmsen, 1966c). It is not known to what extent these substances can re-enter the general metabolic pool. Those pterines (and to a lesser degree purines) which are deposited as dry, irreversible storage mainly in the wings must be regarded as excreted.

Harmsen (1966a) considered the development of dry storage excretion of pterines in Pieridae a possible adaptation for water conservation. A study of the energetics involved in the as yet incompletely understood pathway of pterine synthesis and of the redox potentials of hydrogenated pterines would be of much interest. Even if the total balance of energy would favour uric acid excretion, it would be possible that pterine excretion would allow a more advantageous distribution of oxygen consumption and energy release in relation to the whole process of metamorphosis.

The metabolic connection between purine and pterine metabolic end products may have more biological importance than is understood so far. Both purines and pterines occur in the meconium (Kürsteiner, 1961; Nation, 1963; Harmsen, 1966a; Wessing and Eichelberg, 1968), in the fatbody (Harmsen, 1966a) and in the wings and other parts of the integument (Purrmann, 1940; Harmsen, 1966a, b, c). However, Watt and Bowden (1966) point out that uric acid is present in all species of *Pieridae,* but is not affected by any of the mutants that affect the pterine content. Contrary to this observation is the work of Taira and Nawa (1958) on the mutants *bw* and *W* of *Drosophila* which do not synthesize eye pterines. These mutants lose all uric acid early in adult life, despite the normal level of uric acid in the pupa.

The original concept of pigmentation being the main advantage of pterine deposition (Hopkins, 1895) has been discussed in detail by Harmsen (1966a), who considers pigmentation a secondary advantage. The potential of pterine deposition in the wings, however, may have originally developed in relation to pigmentation. A small xanthopterin containing yellow spot on the forewing of *Vanessa io* (Harmsen, 1963) for instance, could only have a pigmentation advantage. Also, localized, defined deposition of erythropterin as found in *Euchloe, Gonepteryx* and *Mylothris* (see Table II) must be considered primarily as pigmentation: an opinion strengthened by the usual difference in pigmentation between the sexes.

Any further speculation concerning the role of stored simple pterines must obviously await a more detailed understanding of the synthetic pathways involved and of synthetic energetics.

To a certain extent, ommochrome deposition is comparable to pterine storage. These substances can be considered as insoluble, oxidized, non-poisonous end products of tryptophane metabolism (Linzen, 1967). It is possible that in certain insects ommochrome synthesis and deposition (like pterine synthesis and deposition in Pieridae or Dipteran eyes) may have reached a stage where a part of the excretory nitrogen pathway is shunted in this direction. Only quantitative analyses can answer this question. This situation may have developed in Papilionidae where kynurenine and kynurenine derivatives in the wings represent a sizable percentage of the oxidized nitrogen (Umebachi and Katayama, 1966).

XII. CONCLUSION AND GENERAL DISCUSSION

One can visualize an evolutionary sequence starting with the development of hydrogenated pterines from purines or purine precursors. The primary metabolic role must have been, and still is, in their function as co-factors in hydroxylation reactions. The direct oxidation and excretion of the hydrogenated pterines must have replaced their re-entry into purine metabolism for energetic reasons. The auto-chelating properties of simple pterines, resulting in high insolubility under normal physiological conditions or the binding of the eye pterines onto carrier granules, must be the basis of the development of various types of deposition. The storage excretion of insoluble, specialized end products has secondarily developed into a variety of adaptive systems: a much more extensive system of storage excretion of the end products of part of the general nitrogen pool (in Pieridae); a protective coloration (in *Vespa*); a specific and sexual recognition coloration (in Pieridae); a light filtration and screening in the compound eyes (in *Drosophila* and many other insects).

The knowledge and insight of the biological role of pterines in insects exemplifies a pattern of common biological interest: a fundamental physiological process common to all organisms (the biosynthesis of a co-factor from a purine precursor), provides the starting point for a new selection process in physiological development.

REFERENCES

Albert, A. (1954). The pteridines. *Fortschr. Chem. org. NatStoffe* **11**, 350–403.

Almeida, F. F. de (1958). Über die fluoreszierenden Stoffe in den Augen dreier Genotypen von *Plodia interpunctella. Z. Naturf.* **13b**, 687–691.

Aruga, H., Kawase, S. and Akino, M. (1954). Occurrence of an enzyme acting on xanthopterin–B in *Bombyx mori. Experientia* **10**, 336–338.

Auf der Maur, P. (1961). Experimentelle Untersuchungen über den Harnsäurestoffwechsel bei verschiedenen Genotypen von *Drosophila melanogaster. Z. VererbLehre* **92**, 42–62.

Autrum, H. (1955). Die spektrale Empfindlichkeit der Augenmutation *white-apricot* von *Calliphora. Biol. Zbl.* **74**, 515–524.

Autrum, H. (1961). Die Sehschärfe pigmentfreier Facettenaugen von *Calliphora erythrocephala. Biol. Zbl.* **80**, 1–4.

Autrum, H. and Langer, H. (1958). Photolabile Pterine im Auge von *Calliphora erythrocephala. Biol. Zbl.* **77**, 196–201.

Bartel, A. H., Hudson, B. W. and Craig, R. (1958). Pteridines in the milkweed bug, *Oncopeltus fasciatus* (Dallas). I. Identification and localization. *J. Insect Physiol.* **2**, 348–354.

Becker, E. (1937a). Das Fehlen der Pterine in den Excrementen pterinführender Insekten. *Hoppe-Seyler's Z. physiol. Chem.* **246**, 177–180.

Becker, E. (1937b). Über das Pterinpigment bei Insekten und die Färbung und Zeichnung von *Vespa* im besonderen. *Z. Morph. Ökol. Tiere* **32**, 672–751.

Becker, E. and Schöpf, C. (1936). Der mikrochemische Nachweis der Pterine in Insekten. *Justus Liebigs Annln Chem.* **524**, 124–144.

Belzecka, K., Laskowska, T. and Mochnacka, I. (1964). Hydroxylation of phenylalanine in insects and some vertebrates. *Acta biochim. pol.* **11**, 191–196.

Bergman, F. and Kwietny, H. (1959). Pteridines as substrates of mammalian xanthine oxidase. II. Pathways and rates of oxidation. *Biochim. biophys. Acta* **33**, 29–46.

Bernasconi, R. (1963). Sur la présence de ptérine dans les insectes cavernicoles. *Experientia* **19**, 148.

Bhalla, S. C. (1968). Genetic aspects of pteridines in mosquitoes. *Genetics, Princeton* **58**, 249–258.

Boni, P., de Lerma, B. and Parisi, G. (1967). Effects of inhibitor of xanthine dehydrogenase, 4-hydroxypyrazolo (3,4d) pyrimidine (or HPP) on the red eye pigments of *Drosophila melanogaster. Experientia* **23**, 186–187.

Bouthier, A. (1966). Modification des pigments (ommochromes et ptérines) en relation avec la mutation albinos chez, *Locusta migratoria cinerascens* Fabr. *C.r. hebd. Séanc. Acad. Sci., Paris* **262**, 1480–1483.

Brenneman, A. R. and Kaufman, S. (1964). The role of tetrahydropteridines in the enzymatic conversion of tyrosine to 3,4-dihydroxyphenylalanine. *Biochem. biophys. Res. Commun.* **17**, 177–183.

Brenner-Holzach, O. and Leuthardt, F. (1961). Untersuchung über die Biosynthese der Pterine bei *Drosophila melanogaster. Helv. chim. Acta* **44**, 1480–1494.

Brenner-Holzach, O. and Leuthardt, F. (1967). Die Biosynthese der Pterine. IV. Die Herkunft der Seitenkette im Biopterin. *Hoppe-Seyler's Z. physiol. Chem.* **348**, 605–606.

Briggs, M. H. (1961). Retinene–I in insect tissues. *Nature, Lond.* **192**, 874–875.

Burg, A. W. and Brown, G. M. (1968). The biosynthesis of folic acid. VIII. Purification and properties of the enzyme that catalyzes the production of formate from carbon atom 8 of guanosine triphosphate. *J. biol. Chem.* **243**, 2349–2358.

Burgess, L. (1949). Quantitative study of pterin pigment in the developing egg of the grasshopper, *Melanoplus differentialis. Archs Biochem.* **20**, 347–355.

Butenandt, A. (1959). Wirkstoffe des Insektenreiches. *Naturwissenschaften* **46**, 461–471.

Butenandt, A., Biekert, E. and Linzen, B. (1956). Über Ommochrome. VII. Mitt. Modellversuche zur Bildung des Xanthommatins *in vivo. Hoppe-Seyler's Z. physiol. Chem.* **305**, 284–289.

Butenandt, A. and Rembold, H. (1958). Über den Weiselzellenfuttersaft der Honigbiene. II. Isolierung von 2-Amino-4-hydroxy-6-(1,2-dihydroxy-propyl)-pteridin. *Hoppe-Seyler's Z. physiol. Chem.* **311**, 79–83.

Caspari, E. (1955). On the pigment formation in the testis sheath of *Rt* and *rt* *Ephestia kühniella* Zeller. *Biol. Zbl.* **74**, 585–602.

Caspari, E. and Richards, J. (1948). Genic action. *Rep. Carnegie Inst. (Dept. of Genetics)* **47**, 183–189.

Colhoun, E. H. (1963). The physiological significance of acetylcholine in insects and observations upon other pharmacologically active substances. In *Advances in Insect Physiology* (J. W. L. Beament, J. E. Treherne and V. B. Wigglesworth, eds), Vol. 1, pp. 1–46.

Danneel, R. (1955). Über die beiden gelbäugigen Mutanten *se v* und *se cn* der Fruchtfliege. *Naturwissenschaften* **42**, 566.

Davey, K. G. (1964). The control of visceral muscles in insects. In *Advances in Insect Physiology* (J. W. L. Beament, J. E. Treherne and V. B. Wigglesworth, eds), Vol. 2, pp. 219–245.

Descimon, H. (1965a). Identification de la sépiaptérine dans les ailes des Pieridae (Lepidoptera, Rhopalocera). *Bull. Soc. Chim. biol.* **47**, 1095–1100.

Descimon, H. (1965b). Ultrastructure et pigmentation des écailles des Lépidoptères. *J. Microscopie* **4**, 130.

Descimon, H. (1966). Chronologie de la sécrétion des ptérines chez *Colias croceus* Fourcroy (Lepidoptera, Pieridae). *C.r. Séanc. Soc. Biol.* **160**, 928–932.

Descimon, H. (1967). La dihydroxanthoptérine, un nouveau pigment naturel des Lépidoptères. *Bull. Soc. Chim. biol.* **49**, 1164–1166.

Dewey, V. C. and Kidder, G. W. (1966). Effects of long chain mono- and dicarboxylic acids on the pteridine requirement of *Crithidia*. *Archs Biochem. Biophys.* **115**, 401–406.

Dewey, V. C. and Kidder, G. W. (1968). The use of Sephadex for the concentration of pteridines. *J. Chromat.* **31**, 326–336.

Egelhaaf, A. (1956a). Die fluoreszierenden Stoffe in den Organen dreier Genotypen von *Ephestia kühniella*. Beitrag zur Analyse von Genwirkungen im Pterine–und Tryptophanstoffwechsel. *Z. VererbLehre* **87**, 769–783.

Egelhaaf, A. (1956b). Photolabile Fluoreszenzstoffe bei *Ephestia kühniella*. *Naturwissenschaften* **43**, 309.

Ellenbogen, L., Taylor, R. J. and Brundage, G. B. (1965). On the role of pteridines as co-factors for tyrosine hydroxylase. *Biochem. biophys. Res. Commun.* **19**, 708–715.

Euler von, U. S. (1961). Occurrence of catecholamines in acrania and invertebrates. *Nature, Lond.* **190**, 170–171.

Fischer, F., Kapitza, W., Gersch, M. and Unger, H. (1962). Sind Neurohormone der Arthropoden identisch mit fluoreszierenden Substanzen aus dem Nervensystem? *Z. Naturf.* **17b**, 834–836.

Forrest, H. S., Glassman, E. and Mitchell, H. K. (1956). The conversion of 2-amino-4-hydroxypteridine to isoxanthopterin in *Drosophila melanogaster*. *Science, N.Y.* **124**, 725–726.

Forrest, H. S., Hanly, E. W. and Lagowski, J. M. (1961). 2,4-Dihydroxypteridine as an intermediate in the enzymically catalyzed oxidation of 4-hydroxypteridine. *Biochim. biophys. Acta* **50**, 596–598.

Forrest, H. S., Menaker, M. and Alexander, J. (1966). Studies on the pteridines in the milkweed bug, *Oncopeltus fasciatus* (Dallas) *J. Insect Physiol.* **12**, 1411–1421.

Forrest, H. S. and Nawa, S. (1962). Structures of sepiapterin and isosepiapterin. *Nature, Lond.* **196**, 372–373.

Forrest, H. S. and Nawa, S. (1964). Recent work on the structures of isosepiapterin and the drosopterins and its relations to pteridine biosynthesis. In *Pteridine Chemistry* (W. Pfleiderer and E. C. Taylor, eds.) Pergamon Press, London.

Fuge, H. (1966). Über die Bildung der Ommochromgrana von *Drosophila melanogaster*. *Naturwissenschaften* **53**, 136.

Fujimori, E. (1959). Interaction between pteridines and tryptophan. *Proc. natn. Acad. Sci. U.S.A.* **45**, 133–136.

Furneaux, P. J. S. and McFarlane, J. E. (1965). Identification, estimation and localization of catecholamines in eggs of the house cricket, *Acheta domesticus* (L.). *J. Insect Physiol.* **11**, 591–600.

Gal, E. M., Armstrong, J. C. and Ginsberg, B. (1966). The nature of *in vitro* hydroxylation of L-tryptophan by brain tissue. *J. Neurochem.* **13**, 643–654.

Ghosh, D. and Forrest, H. S. (1967a). Enzymatic studies on the hydroxylation of kynurenine in *Drosophila melanogaster*. *Genetics, Princeton* **55**, 423–431.

Ghosh, D. and Forrest, H. S. (1967b). Inhibition of tryptophan pyrrolase by some naturally occurring pteridines. *Archs Biochem. Biophys.* **120**, 578.

Glassman, E. (1962a). Convenient assay of xanthine dehydrogenase in single *Drosophila melanogaster. Science, N.Y.* **137**, 990–991.

Glassman, E. (1962b). *In vitro* complementation between nonallelic *Drosophila* mutants deficient in xanthine dehydrogenase. *Proc. natn. Acad. Sci. U.S.A.* **48**, 1491–1497.

Glassman, E. (1965). Genetic regulation of xanthine dehydrogenase in *Drosophila melanogaster. Fedn Proc. Fedn Am. Socs exp. Biol.* **24**, 1243–1251.

Glassman, E. (1966). Complementation *in vitro* between nonallelic *Drosophila* mutants deficient in xanthine dehydrogenase. III. Observations on heat stabilities. *Biochim. biophys. Acta* **117**, 342–350.

Glassman, E. and Mitchell, H. K. (1959). Mutants of *Drosophila melanogaster* deficient in xanthine dehydrogenase. *Genetics, Princeton* **44**, 153–162.

Goldsmith, T. H. (1958). The visual system of the honey bee. *Proc. natn. Acad. Sci. U.S.A.* **44**, 123–126.

Goldsmith, T. H. and Warner, L. T. (1964). Vitamin A in the vision of insects. *J. gen. Physiol.* **47**, 433–441.

Goodwin, T. W. (1951). Biochemistry of locusts. 6. The occurrence of a flavin in the eggs and of a pterin in the eyes of the African migratory locust (*Locusta migratorioides* R.u.F.) and the desert locust (*Schistocerca gregaria* Forsk). *Biochem. J.* **49**, 84–86.

Goto, M., Okada, T. and Forrest, H. S. (1964). Zur Biosynthese des Drosopterins, eines Augenpigmentes von *D. melanogaster. J. Biochem., Tokyo* **56**, 379d.

Goto, M., Okada, T. and Forrest, H. S. (1965). Synthesis of 2-Amino-4-hydroxy-6-(D-erythro-1′,2′,3′-trihydroxypropyl) pteridine-3′-phosphate-10-C^{14}, and its metabolism in *Drosophila melanogaster. Archs Biochem. Biophys.* **110**, 409–412.

Goto, M., Konishi, M., Sugiura, K. and Tsusue, M. (1966). The structure of a yellow pigment from the mutant lemon of *Bombyx mori. Bull. chem. Soc. Japan* **39**, 929–932.

Gotz, K. G. (1964). Optomotorische Untersuchung des visuellen Systems einiger Augenmutanten der Fruchtfliege *Drosophila. Kybernetik* **2**, 77–92.

Guroff, G. and Strenkoski, C. A. (1966). Biosynthesis of pteridines and of phenylalanine hydroxylase co-factor in cell-free extracts of *Pseudomonas* species (ATCC 11299 a). *J. biol. Chem.* **241**, 2220–2227.

Gutensohn, W. (1968). *Chemische und biochemische Untersuchungen zum Abbau von Tetrahydroneopterin.* Thesis, Ludwig-Max. University, München.

Hackman, R. H. (1964). Chemistry of the insect cuticle. In *The Physiology of Insects* (M. Rockstein, ed.) Vol. III, pp. 471–506. Academic Press, London and New York.

Hadorn, E. and Graf, G. E. (1958). Weitere Untersuchungen über den nicht-autonomen Pterinstoffwechsel der Mutante *rosy* von *Drosophila melanogaster. Zool. Anz.* **160**, 231–243.

Hadorn, E., Graf, G. E. and Ursprung, H. (1958). Der Isoxanthopterin-Gehalt transplantierter Hoden von *Drosophila melanogaster* als nicht-autonomes Merkmal. *Revue suisse Zool.* **65**, 335–342.

Hadorn, E. and Kühn, A. (1953). Chromatographische und fluorometrische Untersuchungen zur biochemischen Polyphänie von Augenfarb-Genen bei *Ephestia kühniella*. *Z. Naturf.* **8b**, 582–589.

Hadorn, E. and Mitchell, H. K. (1951). Properties of mutants of *Drosophila melanogaster* and changes during development as revealed by paper chromatography. *Proc. natn. Acad. Sci. U.S.A.* **37**, 650–665.

Hadorn, E. and Schwinck, I. (1956). Fehlen von Isoxanthopterin und Nicht-Autonomie in der Bildung der roten Augenpigmente bei einer Mutant *(rosy)* von *Drosophila melanogaster*. *Z. VererbLehre* **87**, 528–553.

Hadorn, E. and Ziegler, I. (1958). Untersuchung zur Entwicklung, Geschlechtsspezifität und phänogenetischen Autonomie der Augenpterine verschiedener Genotypen von *Drosophila melanogaster*. *Z. VererbLehre* **89**, 221–234.

Handschin, G. (1961). Entwicklungs- und organspezifisches Verteilungsmuster der Pterine bei einem Wildstamm und bei der Mutante *rosy* von *Drosophila melanogaster*. *Devl Biol.* **3**, 115–139.

Hanser, G. (1948). Über die Histogenese der Augenpigmentgranula bei verschiedenen Rassen von *Ephestia kühniella* Z. und *Ptychopoda seriata* Schrk. *Z. VererbLehre* **82**, 74–97.

Hanser, G. (1959). Genphysiologische Untersuchungen an der *w*-Mutante von *Calliphora erythrocephala*. *Z. Naturf.* **14b**, 194–201.

Hanser, G. and Rembold, H. (1960). Über den Weiselzellenfuttersaft der Honigbiene, IV. Jahreszeitliche Veränderungen im Biopteringehalt des Arbeiterinnenfuttersaftes. *Hoppe-Seyler's Z. physiol. Chem.* **319**, 200–205.

Hanser, G. and Rembold, H. (1964). Analytische und histologische Untersuchungen der Kopf- und Thoraxdrüsen bei der Honigbiene *Apis mellifica*. *Z. Naturf.* **19b**, 938–943.

Hanser, G. and Rembold, H. (1968). Über die gerichtete Aufnahme des Biopterins im Organismus I. Histoautoradiographische Untersuchungen bei der Honigbiene *(Apis mellifica)*. *Z. Naturf.* **23b**, 666–670.

Harmsen, R. (1963). *The storage and excretion of pteridines in Pieris brassicae L. and some other insects*. Thesis, Cambridge University.

Harmsen, R. (1964). Genetically controlled variation in pteridine content of *Pieris brassicae* L. *Nature, Lond.* **204**, 1111.

Harmsen, R. (1966a). The excretory role of pteridines in insects. *J. exp. Biol.* **45**, 1–13.

Harmsen, R. (1966b). Identification of fluorescing and UV absorbing substances in *Pieris brassicae* L. *J. Insect Physiol.* **12**, 23–30.

Harmsen, R. (1966c). A quantitative study of the pteridines in *Pieris brassicae* L. during post-embryonic development. *J. Insect Physiol.* **12**, 9–22.

Harmsen, R. (1969a). The effect of atmospheric oxygen pressure on the biosynthesis of simple pterines in pierid butterflies. *J. Insect Physiol.* (In press.)

Harmsen, R. (1969b). Pteridines in flies of the genus *Glossina* (Diptera). *Acta trop.* (In press.)

Hoffmann, C. and Langer, H. (1961). Die spektrale Augenempfindlichkeit der Mutante *"chalky"* von *Calliphora erythrocephala*. *Naturwissenschaften* **48**, 605.

Hopkins, F. G. (1895). The pigments of Pieridae: a contribution to the study of excretory substances which function in ornament. *Phil. Trans. R. Soc. (B)* **186**, 661–682.

Hubby, J. L. and Throckmorton, L. H. (1960). Evolution and pteridine metabolism in the genus *Drosophila*. *Proc. natn. Acad. Sci. U.S.A.* **46**, 65–78.

Hudson, B. W., Bartel, A. H. and Craig, R. (1959). Pteridines in the milkweed bug, *Oncopeltus fasciatus* (Dallas) II. Quantitative determination of pteridine content of tissues during growth. *J. Insect Physiol.* **3**, 63–73.

Ikan, R. and Ishay, J. (1967). Pteridines and purines of the queens of the oriental hornet, *Vespa orientalis* F. *J. Insect Physiol.* **13**, 159–162.

Ikeda, M., Levitt, M. and Udenfried, S. (1967). Phenylalanine as substrate and inhibitor of tyrosine hydroxylase. *Archs Biochem. Biophys.* **120**, 420–427.

Jaenicke, L. and Kutzbach, C. (1963). Folsäure und Folat-Enzyme. In *Fortschritte der Chemie organischer Naturstoffe* (L. Zechmeister, ed.). Vol. XXI, pp. 184–227. Springer, Vienna.

Jones, R. R. (1968). Personal communication, *Cambridge*.

Kalmus, H. (1943). The optomotor responses of some eye mutants of *Drosophila*. *J. gen. Physiol.* **45**, 206–213.

Katoh, S. and Akino, M. (1966). *In vitro* conversion of sepiapterin to isosepiapterin via dihydrobiopterin. *Experientia* **22**, 793.

Kaufman, S. (1958). A new co-factor required for the enzymatic conversion of phenylalanine to tyrosine. *J. biol. Chem.* **230**, 931–939.

Kaufman, S. (1959). Studies on the mechanism of the enzymatic conversion of phenylalanine to tyrosine. *J. biol. Chem.* **234**, 2677–2682 and 2683–2686.

Kaufman, S. (1962). On the structure of phenylalanine hydroxylation co-factor *J. biol. Chem.* **237**, PC 2712–2713.

Kaufman, S. (1963). The structure of the phenylalanine-hydroxylating co-factor *Proc. natn. Acad. Sci. U.S.A.* **50**, 1085–1093.

Kaufman, S. (1967). Pteridine co-factors. *A. Rev. Biochem.* **36**, 171–184.

Keller, E. C. and Glassman, E. (1965). Phenocopies of the *ma-l* and *ry* mutants of *Drosophila melanogaster*: inhibition *in vivo* of xanthine dehydrogenase by 4-hydroxy-pyrazolo (3,4-d) pyrimidine. *Nature, Lond.* **208**, 202–203.

Kidder, G. W. and Dewey, V. C. (1963). Relationship between pyrimidine and lipid biosynthesis and unconjugated pteridine. *Biochem. biophys. Res. Commun.* **12**, 280–283.

Kishi, Y., Matusuura, S., Inoue, S., Shimonura, O. and Goto, T. (1968). Luciferin and luciopterin isolated from the Japanese firefly, *Luciola cruciata*. *Tetrahedron Lett.* **24**, 2847–2850.

Köhler, F. (1958). Papierchromatographische Untersuchungen fluoreszierender Stoffe der Honigbiene. *Naturwissenschaften* **45**, 421–422.

Kokolis, N. and Ziegler, I. (1968). Wiedererscheinen von Tetrahydrobiopterin in der Regenerationsknospe von *Triturus*-Arten. *Z. Naturf.* **23b**, 860–865.

Krebs, E. G. and Norris, E. R. (1949). The competitive inhibition of xanthine oxidation by xanthopterin. *Archs Biochem.* **24**, 49–54.

Krieger, F. (1954). Untersuchungen über den Farbwechsel der Libellenlarven. *Z. vergl. Physiol.* **36**, 352–366.

Kühn, A. (1955). Ein Fluorometer für Papierchromatogramme. *Naturwissenschaften* **42**, 529–530.

Kühn, A. and Almeida, F. de (1961). Fluoreszierende Stoffe und Ommochrome bei Genotypen von *Plodia interpunctella*. *Z. VererbLehre* **92**, 126–132.

Kühn, A. and Berg, B. (1955). Zur genetischen Analyse der Mutation *biochemica* von *Ephestia kühniella. Z. VererbLehre* 87, 25–35.

Kühn, A. and Berg, B. (1956). Über die Kombination von *wa* (weissäugig) und *biochemica* bei *Ephestia kühniella. Z. VererbLehre* 87, 335–337.

Kühn, A. and Egelhaaf, A. (1955). Zur chemischen Polyphänie bei *Ephestia kühniella. Naturwissenschaften* 42, 634–635.

Kühn, A. and Egelhaaf, A. (1959). Der rote Augenfarbstoff von *Ephestia* und *Ptychopoda* ein Pterinpigment. *Z. Naturf.* 14b, 654–659.

Kürsteiner, R. (1961). Über die fluoreszierenden Stoffe (Pterine) in den Meconien der Wildrasse und der Mutanten *white* und *rosy*[2] von *Drosophila melanogaster. J. Insect Physiol.* 7, 5–31.

Langer, H. (1967). Über die Pigmentgranula im Facettenauge von *Calliphora erythrocephala. Z. vergl. Physiol.* 55, 354–377.

Langer, H. and Hoffmann, C. (1966). Elektro- und stoffwechsel-physiologische Untersuchungen über den Einfluss von Ommochromen und Pteridinen auf die Funktion des Facettenauges von *Calliphora erythrocephala. J. Insect Physiol.* 12, 357–387.

Levenberg, B. and Hayaishi, O. (1959). A bacterial deaminase. *J. biol. Chem.* 234, 955–961.

Levenberg, B. and Kaczmarek, D. K. (1966). Enzymic release of carbon atom 8 from guanosine triphosphate, an early reaction in the conversion of purines to pteridines. *Biochim. biophys. Acta* 117, 272–275.

L'Hélias, C. (1956). Les hormones du complexe rétrocérébrale du phasme *Carausius morosus:* action chimique et identification du squelette commun de ces hormones. *Annls Biol.* 32, 203–219.

L'Hélias, C. (1957). Isolement de substances pré- ou co-hormonales du complexe postcérébrale de *Carausius morosus* et de *Clitumnus extradentatus. Bull. biol. Fr. Belg.* 91, 241–263.

L'Hélias, C. (1961). Corrélations entre les ptérines et le photoperiodisme dans la régulation du cycle sexual chez les Pucerons. *C.r. hebd. Séanc. Acad. Sci. Paris* 253, 1353–1355.

L'Hélias, C. (1962). Rétablissment du cycle bisexual par la bioptérine et l'acide folique chez le puceron parthénogénétique *Sappaphis plataginea. C.r. hebd. Séanc. Acad. Sci. Paris* 255, 388–390.

Linzen, B. (1967). Zur Biochemie der Ommochrome. Unterteilung, Vorkommen, Biosynthese und physiologische Zusammenhänge. *Naturwissenschaften* 54, 259–267.

Linzen, B. and Hertel, U. (1967). *In vitro* assay of insect kynurenine-3-hydroxylase. *Naturwissenschaften* 54, 21.

Lovenberg, W., Jequier, E. and Sjoerdesma, A. (1967). Tryptophan hydroxylation: measurement in Pineal gland, Brainstem, and Carcinoid Tumor. *Science, N.Y.* 155, 217–219.

Maier, W. (1965). Elektronenmikroskopischer Vergleich von Retinula-Pigmentzellen der Wildfrom *wa*[+] und der Mutante *wa* von *Ephestia kühniella. Z. Naturf.* 20b, 312–314.

Mason, S. F. (1954). Some aspects of the ultraviolet spectra of the pteridines. In *Chemistry and Biology of Pteridines* (G. E. W. Wolstenholme and M. P. Cameron, eds) Little, Brown and Co., Boston.

Mathews, C. K., Scrimegeour, R. G. and Huennekens, F. M. (1963). In *Methods of Enzymology* (S. P. Colowick and N. O. Kaplan, eds) Vol. 6, 364 pp. Academic Press, New York.

Matsubara, M. and Akino, M. (1964). On the presence of sepiapterin reductase different from folate and dihydrofolate reductase in chicken liver. *Experientia* 20, 574–575.

Matsubara, M., Katoh, S., Akino, M. and Kaufman, S. (1966). Sepiapterin reductase. *Biochim. biophys. Acta* 122, 202–212.

Matsubara, M., Tsusue, M. and Akino, M. (1963). Occurrence of two different enzymes in the silkworm, *Bombyx mori*, to reduce folate and sepiapterin. *Nature, Lond.* 199, 908–909.

McLean, F. I., Forrest, H. S. and Myers, J. (1965). Origin of the sidechain in pteridines of the biopterin type. *Biochem. biophys. Res. Commun.* 18, 623–626.

McNutt, W. S. (1956). The incorporation of the pyrimidine ring of adenine into the isoalloxazine ring of riboflavin. *J. biol. Chem.* 219, 365–373.

McNutt, W. S. (1963). The metabolism of isoxanthopterin by *Alcaligenes faecalis. J. biol. Chem.* 238, 1116–1121.

McNutt, W. S. (1964). Tetraoxypteridine isomerase. *J. biol. Chem.* 239, 4272–4279.

Merlini, L. and Mondelli, R. (1962). Sui pigmenti di *Pyrrhocoris apterus* L. *Gazz. chim. ital.* 92, 1251–1261.

Merlini, L. and Nasini, G. (1966). Insect Pigments. IV. Pteridines and colour in some Hemiptera. *J. Insect Physiol.* 12, 123–127.

Müller, G. (1956). Fluoreszierende Stoffe in *Vaness io* L. chromatographisch untersucht. *Z. Naturf.* 11b, 221–222.

Munz, P. (1964). Untersuchungen über die Aktivität der Xanthindehydrogenase in Organen und während der Ontogenese von *Drosophila melanogaster. Z. VererbLehre* 95, 195–210.

Muth, F. W. (1967). Quantitative Untersuchungen zum Pigmentierungsgeschehen in den Augen der Mehlmotte *Ephestia kühniella. Wilhelm Roux Arch. EntwMech. Org.* 159, 379–411.

Nathan-Guttman, H. A. (1964). Crithidia assays of unconjugated pteridines. In *Pteridine Chemistry.* (W. Pfleiderer and E. C. Taylor, eds) Pergamon Press, London.

Nathan, H. A. and Ziegler, I. (1961). Microbial incubation experiments with the eye pteridines of *Drosophila melanogaster. Z. Naturf.* 16b, 262–264.

Nation, J. L. (1963). Identification of xanthine in excreta of the greater wax moth, *Galleria mellonella* (L.), *J. Insect Physiol.* 9, 195–200.

Nawa, S., Sakaguchi, B. and Taira, T. (1957). Genetical and biochemical studies on the metabolism of pteridines in insects. *Rep. natn. Inst. Genet. Misima* 7, 32–35.

Nolte, D. J. (1961). Submicroscopic structure of *Drosophila* eye. *Afr. J. Sci.* 57, 121–125.

Ohnishi, E. (1959). Pigment composition in the pupal cuticles of two colour types of the swallowtails, *Papilio xuthus* L. and *P. protenor demetrius* Cramer. *J. Insect Physiol.* 3, 132–145.

Okada, T. and Goto, M. (1965). Synthesis of 2-amino-4-hydroxy-6-hydroxy methylpteridine-10-C[14] and their metabolism in *Drosophila melanogaster. J. Biochem.* 58, 458–462.

Osborne, M. J. and Huennekens, F. M. (1958). Enzymatic reduction of dihydro-folic acid. *J. biol. Chem.* **233**, 969–974.
Patat, U. (1956). Über das Pterinmuster der Facettenaugen von *Calliphora erythrocephala*. Ein Beitrag zur Funktion und Stabilität der Pterine. *Z. vergl. Physiol.* **51**, 103–134.
Patterson, E. C., Milstrey, R. and Stockstad, E. C. (1958). The synthesis of some 2-amino-4-hydroxy-6-polyhydroxyalkylpteridines, which are active in supporting the growth of the protozoon *Crithidia fasciculata*. *J. Am. chem. Soc.* **80**, 2018–2020.
Pfleiderer, W. (1962). Pteridine. XXIV. Über die Isolierung und Struktur des orangeroten Schmetterlingspigmentes "Erythropterin". *Chem. Ber.* **95**, 2195–2204.
Pfleiderer, W. (1963). Neurere Entwicklungen in der Pteridin-Chemie. *Angew. Chem.* **75**, 993–1011.
Pfleiderer, W. (1964). Neues über natürliche Pteridine. *Angew. Chem.* **76**, 757.
Plaut, G. W. E. (1964). Enzymic formation of riboflavin. In *Pteridine Chemistry* (W. Pfleiderer and E. C. Taylor eds) Pergamon Press, London.
Polonovski, M. and Busnel, R. G. (1948). Physiologie biochemique sur les pigments des oeufs de *Bombyx mori*. VII. *Congr. ser. intern. Ales-France 1948*, pp. 621–622.
Purrmann, R. (1940). Über die Flügelpigmente der Schmetterlinge. VII. Synthese des Leukopterins und Natur des Guanopterins. *Justus Liebigs Annln Chem.* **544**, 182–190.
Reisener-Glasewald, E. (1956). Über die Entwicklung des Bestandes an fluores-zierenden Stoffen in den Köpfen von *Ephestia kühniella* in Abhängigkeit von verschiedenen Augenfarbgenen. *Z. VererbLehre* **87**, 668–693.
Rembold, H. and Buschmann, L. (1962). Trennung von 2-Amino-4-hydroxy-pteridinen durch Ionenaustauscher–Chromatographie. *Hoppe-Seyler's Z. physiol. Chem.* **330**, 132–139.
Rembold, H. and Buschmann, L. (1963a). Untersuchungen über die Pteridine der Bienenpuppe *(Apis mellifica)*. *Justus Liebigs Annln Chem.* **662**, 72–82.
Rembold, H. and Buschmann, L. (1963b). Struktur und Synthese des Neo-pterins. *Justus Liebigs Annln Chem.* **662**, 1406–1410.
Rembold, H. and Gutensohn, W. (1968). 6-Hydroxylation of the pteridine ring by xanthine oxidase. *Biochem. biophys. Res. Commun.* **31**, 837–841.
Rembold, H. and Hanser, G. (1960a). Über den Weiselzellenfuttersaft der Honigbiene. VI. Der Stoffwechsel des Biopterins in der Honigbiene. *Hoppe-Seyler's Z. physiol. Chem.* **319**, 213–219.
Rembold, H. and Hanser, G. (1960b). Über den Weiselzellenfuttersaft der Honigbiene. V. Untersuchungen über die Bildung des Futtersaftes in der Ammenbiene. *Hoppe-Seyler's Z. physiol. Chem.* **319**, 206–212.
Rembold, H. and Hanser, G. (1964). Über den Weiselzellenfuttersaft der Honigbiene. VIII. Nachweis des determinierenden Prinzips im Futtersaft der Königinnenlarven. *Hoppe-Seyler's Z. physiol. Chem.* 251–254.
Rembold, H. and Simmersbach, F. (1969). Catabolism of Pteridine Cofactors, II. A specific pterine deaminase in rat liver. *Biochim. biophys. Acta* (In press).
Rembold, H., Metzger, H., Sudershan, P. and Gutensohn, W. (1969). Catabolism of Pterine Cofactors, I. Properties and Metabolism in Rat Liver Homo-genates of Tetrahydro-biopterin and -neopterin. *Biochim. biophys. Acta* (In press).

Reynolds, J. J. and Brown, G. M. (1964). The biosynthesis of folic acid. IV. Enzymatic synthesis of dihydrofolic acid from guanine and ribose compounds. *J. biol. Chem.* **239**, 317–325.

Sakaguchi, B. (1955). Biochemical and genetical studies on wild silkworm. II. On the nature of the pigments in the epidermal tissues of the Chinese Tussar silkworm, *Antheraea pernyi. Rep. natn Inst. Genet., Misima* **5**, 1955.

Saul, G. B. (1960). The occurrence of fluorescent substances in the parasitic wasp *Mormoniella vitripennis. Revue suisse Zool.* **67**, 270–281.

Schmidt, G. H. and Viscontini, M. (1962). Fluoreszierende Stoffe aus Roten Waldameisen der Gattung *Formica* (Ins., Hym.). I. Isolierung von Riboflavin, 2-Amino-6-hydroxypteridin, Isoxanthopterin, Biopterin und einer neuen, als "Formicapterin" bezeichneten Substanz. *Helv. chim. Acta* **45**, 1571–1575.

Schmidt, G. H. and Viscontini, M. (1964). Fluoreszierende Stoffe aus Roten Waldameisen der Gattung *Formica*. II. Isolierung einer Riboflavin-Formicapterin-Verbindung. *Helv. chim. Acta* **47**, 2049–2052.

Schmidt, G. H. and Viscontini, M. (1966). Fluoreszierende Stoffe aus Roten Waldameisen der Gattung *Formica* (Ins., Hym.). III. Isolierung von D-oder L-6-(threo-1′,2′,3′-trihydroxypropylpterin und Pterincarbonsäure aus Ameisenmännchen. *Helv. chim. Acta* **49**, 344–349.

Schmidt, G. H. and Viscontini, M. (1967). Fluoreszierende Stoffe aus Roten Waldameisen der Gattung *Formica* (Ins., Hym.). V. Isolierung von Lumazin-Derivaten aus Ameisenmänchen. *Helv. chim. Acta* **50**, 34–42.

Schmidt, K. (1965). Beiträge zur Biogenese des Leukopterins. Biogenetischer Zusammenhang zwischen Purinnucleotiden und Pterinen beim grossen Kohlweissling. Thesis, Lugwig-Max. University, München.

Schöpf, C. (1964). Die Anfänge der Pterinchemie. In *Pteridine Chemistry* (W. Pfleiderer and E. C. Taylor eds) Pergamon Press, London.

Schöpf, C. and Becker, E. (1933). Uber das Vorkommen der Pterine in Wespen und Schmetterlingen und über einige neue Beobachtungen am Leukopterin und Xanthopterin. *Justus Liebigs Annln Chem.* **507**, 266–296.

Schöpf, C. and Becker, E. (1936). Über neue Pterine. *Justus Liebigs Annln Chem.* **524**, 49–123.

Schöpf, C. and Reichert, R. (1941). Zur Kenntnis des Leukopterins. *Justus Liebigs Annln Chem.* **548**, 82–94.

Schöpf, C. and Weiland, H. (1926). Über das Leukopterin, das weisse Flügel-pigment der Kohlweisslinge *(Pieris brassicae* und *P. napi). Ber. dt. chem. Ges.* **59**, 2067–2072.

Schwinck, I. (1965). Experimentelle Beeinflussung der Drosopterinsynthese in den *Drosophila*–Mutanten *rosy* und *maroon-like. Z. Naturf.* **20b**, 322–326.

Shoup, J. R. (1966). The development of pigment granules in the eyes of wild type and mutant *Drosophila melanogaster. J. biophys. biochem. Cytol.* **29**, 223–249.

Simon, H., Weygand, F., Walter, J., Wacker, H. and Schmidt, K. (1963). Zusammenhänge zwischen Purin- und Leucopterin-Biogenese in *Pieris brassicae* L. *Z. Naturf.* **18b**, 757–764.

Strehler, B. L. (1951). The isolation and properties of firefly luciferescein. *Archs Biochem.* **32**, 397–406.

Sugiura, K. and Goto, T. (1967). Biosynthesis of pteridines in *D. melanogaster*. *Biochem. biophys. Res. Commun.* **28**, 687–691.

Taira, T. (1961a). Enzymatic reduction of the yellow pigment of *Drosophila*. *Nature, Lond.* **189**, 231–232.

Taira, T. (1961b). The metabolism of sepiapterin in *Drosophila melanogaster*, emphasizing its tetrahydro-form. *Jap. J. Genet.* **36**, 244–256.

Taira, T. (1961c). Comparisons of pterine reductase activity between eye-colour mutants of *Drosophila melanogaster*. *Jap. J. Genet.* **36**, 210–211.

Taira, T. and Nawa, S. (1958). No direct metabolic relation between pterines and uric acid, flavins or folic acid in *Drosophila melanogaster*. *Jap. J. Genet.* **33**, 42–45.

Tartter, A. (1940). Harnsäure und Hypoxanthin als Pigmentbestandteile der Flügel von Pieriden. *Hoppe-Seyler's Z. physiol. Chem.* **266**, 130–134.

Tate, E. (1947). The effect of cold upon the development of pigment in a white-eyed mutant of the blow-fly *(Calliphora erythrocephala)*. *J. Genet.* **48**, 192–193.

Tschesche, R. and Korte, F. (1951). Über Pteridine; IV. Mitt.: Zur Konstitution des Chrysopterins und Mesopterins. *Chem. Ber.* **84**, 641–648.

Tsujita, M. (1961). Maternal effect of $^{d\text{-}lem}$ gene on pterine reductase of *Bombyx mori, Jap. J. Genet.* **36**, 337–346.

Tsujita, M. and Sakaguchi, B. (1955). Genetical and biochemical studies of yellow lethal larvae in the silkworm. (1). On the nature of pterine obtained from the yellow lethal strain. *Jap. J. Genet.* **30**, 83–88.

Tsujita, M. and Sakurai, S. (1963). The association of a specific protein with yellow pigments (Dihydropterin) in the silkworm, *Bombyx mori. Jap. J. Genet.* **38**, 97–105.

Tsusue, M. (1967). Occurrence of sepiapterin deaminase in the silkworm, *Bombyx mori. Experientia* **23**, 116–117.

Umebachi, Y. and Katayama, M. (1966). Tryptophan and tyrosin metabolism in the pupae of Papilionid butterflies. II. *J. Insect Physiol.* **12**, 1539–1547.

Ursprung, H. and Hadorn, E. (1961). Xanthindehydrogenase in Organen von *Drosophila melanogaster. Experientia* **17**, 230.

Viscontini, M. (1958). Fluoreszierende Stoffe aus *Drosophila melanogaster*. 9. Mitt. Kristallisiertes Isodrosopterin. *Helv. chim. Acta* **41**, 922–924.

Viscontini, M. (1964). Betrachtungen über Konstitutionen von Sepiapterinen und Drosopterinen. In *Pteridine Chemistry* (W. Pfleiderer and E. C. Taylor eds) Pergamon Press, London.

Viscontini, M., Hadorn, E. and Karrer, P. (1957). Fluoreszierende Stoffe aus *Drosophila melanogaster:* die roten Augenfarbstoffe. *Helv. chim. Acta* **40**, 579–585.

Viscontini, M., Kühn, A. and Egelhaaf, A. (1956). Isolierung Fluoreszierender Stoffe aus *Ephestia kühniella. Z. Naturf.* **11b**, 501–504.

Viscontini, M. and Möhlmann, E. (1959a). Fluoreszierende Stoffe aus *Drosophila melanogaster*. 12. Mitt. Die gelb fluoreszierende Pterine Sepiapterin und Isosepiapterin. *Helv. chim. Acta* **42**, 836–841.

Viscontini, M. and Möhlmann, E. (1959b). Fluoreszierende Stoffe aus *Drosophila melanogaster*. 13. Mitt. Weitere Beiträge zur Konstitutionsaufklärung der Sepiapterine und der Drosopterine. *Helv. chim. Acta* **42**, 1679–1683.

202   I. ZIEGLER AND R. HARMSEN

Viscontini, M. and Schmidt, G. H. (1963). Fluoreszierende Stoffe aus *Rhodnius prolixus* Stal (Hemiptera; Heteroptera). *Helv. chim. Acta* **46**, 2509–2516.
Viscontini, M. and Schmidt, G. H. (1966). Fluoreszierende Stoffe aus Roten Waldameisen der Gattung *Formica* (Ins., Hym.). 4. Mitt. Isolierung von Isoalloxazin-Derivaten aus Ameisenmännchen. *Helv. chim. Acta* **49**, 1259–1265.
Viscontini, M. and Stierlin, H. (1962a). Fluoreszierende Stoffe aus *Ephestia kühniella* Zeller. III. Isolierung und Strukturen von Erythropterin, Ekapterin und Lepidopterin. *Helv. chim. Acta* **45**, 2479–2487.
Viscontini, M. and Stierlin, H. (1962b). Fluoreszierende Stoffe aus *Ephestia kühniella* Zeller. 4. Mitt. Synthese von Erythropterin, Ekapterin und Lepidopterin. *Helv. chim. Acta* **46**, 51–56.
Viscontini, M. and Weilenmann, H. R. (1959). Über Pteridinchemie. 2. Mitt. Rückoxydation des 2-Amino-6-hydroxy-7,8,9,10-tetrahydropterins an der Luft. *Helv. chim. Acta* **42**, 1854–1862.
Wacker, H., Harvey, R. A., Winestock, C. H. and Plaut, G. W. E. (1964). 4-(1'-D-Ribitylamino)-5-amino-2,6-dihydroxy-pyrimidine, the second product of the riboflavin synthetase reaction. *J. biol. Chem.* **239**, 3493–3497.
Watt, W. B. (1964). Pteridine components of wing pigmentation in the butterfly *Colias eurytheme*. *Nature, Lond.* **201**, 1326–1327.
Watt, W. B. (1967). Pteridine biosynthesis in the butterfly *Colias eurytheme*. *J. biol. Chem.* **242**, 565–572.
Watt, W. B. and Bowden, S. R. (1966). Chemical phenotypes of pteridine colour forms in *Pieris* butterflies. *Nature, Lond.* **210**, 304–306.
Weber, F. and Wiss, O. (1966). In *Hoppe-Seyler-Thierfelder: Handbuch der Physiologisch–und Pathologisch–Chemischen Analyse* 10th edit. Vol. VIB, p. 848. Springer, Berlin, Heidelberg, New York.
Wessing, A. and Eichelberg, D. (1968). Die fluoreszierenden Stoffe aus den Malpighischen-Gefässen der Wildform und verschiedener Augenfarben-mutanten von *Drosophila melanogaster*. *Z. Naturf.* **23b**, 376–386.
Weygand, F., Simon, H., Dahms, G., Waldschmidt, M., Schliep, H. J. and Wacker, H. (1961). Über die Biogenese des Leukopterins. *Angew. Chem.* **73**, 402–407.
Wieland, H. and Liebig, R. (1944). Ergänzende Beiträge zur Kenntnis der Pteridine. *Justus Liebigs Annln Chem.* **555**, 146–156.
Wieland, H. and Schöpf, C. (1925). Über den gelben Flügelfarbstoff des Citronenfalters *(Gonepteryx rhamni)*. *Ber. dt. chem. Ges.* **58**, 2178–2183.
Zeutzschel, B. (1958). Entwicklung und Lage der Augenpigmente bei verschie-denen *Drosophila*-Mutanten. *Z. VererbLehre* **89**, 508–520.
Ziegler, I. (1960a). Tetrahydrobiopterin-Derivat als lichtempfindliche Verbindung bei Amphibien und Insekten. *Z. Naturf.* **15b**, 460–465.
Ziegler, I. (1960b). Zur Feinstruktur der Augengranula bei *Drosophila melano-gaster*. *Z. VererbLehre* **91**, 206–209.
Ziegler, I. (1961a). Zur genphysiologischen Analyse der Pterine in Insektenaugen *(Drosophila melanogaster* und *Calliphora erythrocephala)*. *Z. VererbLehre* **92**, 239–245.
Ziegler, I. (1961b). Genetic aspects of ommochromes and pterine pigments. *Adv. Genet.* **10**, 349–403.

Ziegler, I. (1963a). Reinigung des Tetrahydropterins aus Insektenaugen. *Biochim. biophys. Acta* **78**, 219–220.

Ziegler, I. (1963b). Tetrahydropterin und Melanophoren-Differenzierung bei Fischen. *Z. Naturf.* **18b**, 551–556.

Ziegler, I. (1965). Pterine als Wirkstoffe und Pigmente. *Ergebn. Physiol.* **56**, 1–66.

Ziegler, I. and Feron, M. (1965). Quantitative Bestimmung des hydrierten Pterine und des Xanthommatins in den Augen von *Ceratitis capitata* Wild (Dipt., Trypetidae) *Z. Naturf.* **20b**, 318–322.

Ziegler, I. and Hadorn, E. (1958). Manifestation rezessiver Augenfarbgene im Pterin-Inventar Heterozygoter Genotypen von *Drosophila melanogaster. Z. VererbLehre* **89**, 235–245.

Ziegler, I. and Jaenicke, L. (1959). Zur Wirkungsweise des *white*-Allels bei *Drosophila melanogaster. Z. VererbLehre* **90**, 53–61.

Ziegler, I. and Nathan, H. A. (1961). Wuchsstoffaktivität der Augenpterine von *Drosophila melanogaster* bei *Crithidia fasciculata. Z. Naturf.* **16b**, 260–262.

Electrochemistry of Insect Muscle

P. N. R. USHERWOOD

*Department of Zoology,
University of Glasgow, Glasgow*

I. INTRODUCTION

Insect muscles, although always striated, can be broadly classified into skeletal, cardiac and visceral muscles. The skeletal muscles originate and insert on skeletal structures (i.e. endoskeleton or exoskeleton) while the visceral muscles invest the internal organs and lack strict origins and insertions. The majority of insect skeletal muscles are phasically responsive (i.e. they contract and relax very quickly) the total duration of the isometric twitch contraction being always less than 1 sec and usually less than 100 msec (Usherwood,

1962a). However, some skeletal muscles contain fibres which have much longer contraction cycles than this and can be described as tonically responsive (Usherwood, 1967b; Cochrane *et al.,* 1969). Insect visceral muscles, which include muscles of the reproductive system and of the gut are mainly tonically responsive and like cardiac muscle often contract rhythmically.

A number of different types of insect skeletal muscle fibre have been described. These fibres differ in the number and disposition of their nuclei and mitochondria, in the arrangement and amount of sarcoplasmic reticulum (SR) and sarcoplasm, and in the structure of their myofibrils (Wigglesworth, 1965; Smith, 1966; Usherwood, 1967b). However, they all have certain features in common. For example, they are characterized by an outer limiting membrane, the sarcolemma, enclosing the contents of the fibres, which include the mitochondria, sarcoplasm, sarcoplasmic reticulum, nuclei and the fibrils containing the contractile elements. The latter consist of the familiar interdigiting thick (myosin?) and thin (actin?) filaments characteristic of vertebrate skeletal muscle. Insect skeletal muscle fibres are always regularly striated. They are divided transversely into sarcomeres by dense lines, the Z-lines. The striation pattern (A-bands, I-bands, Z-lines) of the fibres seen when insect skeletal muscle fibres are examined, using light microscopy and electron microscopy, is a reflection of this regular transverse division together with the regular arrangement within each sarcomere of the myosin and actin filaments. In the larger insects the muscle fibres are usually about 100 μ in diameter although some fibrillar flight muscle fibres may reach a diameter of up to 1.8 mm e.g. in the tachinid fly, *Rutilia potina* (Diptera) (Pringle, 1957).

Insect skeletal muscles are innervated by two main classes of motoneurone, excitatory and inhibitory (Usherwood, 1967b). The fibres of some muscles receive endings from more than one axon (i.e. they are polyneuronally innervated). The nature of the nervous innervation is of considerable importance in terms of the electro-chemistry of insect skeletal muscle fibres for it will be shown later that the ionic basis for electrogenesis at inhibitory synapses is strikingly different from that at excitatory synapses. Presumably development of different nervous inputs has involved the evolution of differentiated regions of muscle membrane with specific pharma-cological and ionic properties.

Smith *et al.* (1966) studied the ultrastructure of three visceral muscle systems; the muscular sheath investing the midgut of the larvae of a moth, *Ephestia sp.* (Lepidoptera), the spermathecae of a cockroach, *Periplaneta sp.* (Dictyoptera) and the seminal vesicle of a stick insect, *Carausius sp.* (Cheleutoptera). The muscle fibres surrounding the seminal vesicle of *Carausius* are disposed in both a longitudinal and a circular fashion. Elsewhere the visceral fibres may form an irregular feltwork of branching and anastomizing fibres (Morrison, 1928). The contractile system of each of these visceral fibres corresponds essentially to a single myofibril. Each fibre has well defined Z-lines. In the seminal vesicle of *Carausius* the sarcomere length of the resting muscle fibre is 7–8 μ (Smith *et al.,* 1966). In gut muscles and muscles of the reproductive system, the contractile material consists of a double array of thick and thin filaments, a configuration not unlike that seen in some insect leg muscles (Usherwood, 1967b; Cochrane *et al.,* 1969) but strikingly different from that of insect flight muscle fibres (Smith, 1965). The muscle fibres of the reproductive system and the gut are never greater than 26 μ in width (Smith *et al.,* 1966; Belton and Brown, 1969) and are usually uninucleate (Smith *et al.,* 1966).

Insect cardiac muscle has a micromorphology more reminiscent of insect gut muscle and muscle of the insect reproductive system than of vertebrate cardiac muscle. However, structures similar in appearance to the intercalated discs found in vertebrate heart muscle have been seen in some insect hearts (Edwards and Challice, 1960; McCann, 1965; Sanger and McCann, 1968). In the heart of the moth, *Hyalophora cercropia* (Lepidoptera) the muscle fibres are arranged in a non-orderly feltwork of two layers (McCann, 1965) whereas the heart muscle of *Periplaneta americana* is only one cell thick (Edwards and Challice, 1960). Insect cardiac muscle fibres are about 30 μ in diameter and 100 μ long (McCann, 1965) and multinucleate (Sanger and McCann, 1968); the arrangement of the contractile elements, sarcoplasm, sarcoplasmic reticulum and mitochondria is similar to that seen in insect gut muscle.

Information on the innervation of insect visceral and cardiac muscles and on the nervous and endocrine control of activity of these muscles is quite extensive (Davey, 1964; Jones, 1964) but, as yet, rather inconclusive. Muscles of the ventral diaphragm or septum which separates the perivisceral sinus from the perineural sinus have been classified as part of the visceral musculature (Davey, 1964). Guthrie (1962) examined some of the properties of the cockroach ventral diaphragm and demonstrated motor and sensory innervation of this structure. Apart from this, the properties of the ventral diaphragm and, for that matter, the dorsal diaphragm also, remain obscure.

II. THE MUSCLE MEMBRANE

A. STRUCTURE AND MORPHOLOGY

When viewed with the light microscope, insect muscle fibres appear to be clearly bounded by a simple limiting membrane. However, this "membrane", which has been called the sarcolemma, is a complex structure, and when viewed with the electron microscope, is seen to consist of a cell membrane proper, or plasma membrane, closely associated with a diffuse structure, the basement membrane, or external lamina (Fawcett, 1966), which lies outside it.

In electron micrographs of insect skeletal and visceral muscles the plasma membrane is usually about 7.5 mμ in width and exhibits the triple-layered "unit membrane" organization characteristic of other cell membranes. The triple-layered structure is most clearly seen after fixation with potassium permanganate although osmium fixed membranes exhibit this organization to some extent. Unfortunately, this is the limit of our knowledge of the structural properties of insect muscle membranes. Permeability studies, X-ray diffraction studies, polarized light studies, histochemical studies and studies of physical parameters that have been made on a variety of animal cell membranes to elucidate their structural properties, have not been repeated so far for insect muscle membranes. In general, most of the cell membranes that have been studied using these techniques show a high permeability for lipid soluble substances, a low surface tension and a high electrical resistance. It seems probable that these membranes contain a continuous lipid layer made up of phosphatide, sterol and fat molecules and that the surface of the membrane is covered by adsorbed proteins (Brown and Danielli, 1964). Robertson (1960) explains the triple-layered "unit membrane" structure, seen in electron micrographs of cell membranes, on the basis that the dense outer lines represent the polar ends of the lipid molecules arranged in a bilayer, whilst the clear central zone represents non-polar unsaturated carbon chains. Since our knowledge of insect muscle membranes is so limited a more advanced review of current ideas on membrane structure seems inappropriate here.

The plasma membrane of insect muscle fibres is invested by an extracellular layer of basement membrane. This basement membrane appears in electron micrographs as a finely granular or amorphous matrix. In skeletal muscle fibres the basement membrane usually appears to be devoid of fibrous material (Ashhurst, 1968), although collagen-like fibres have been found infrequently in the basement membrane of locust leg muscle fibres (Cochrane, unpublished) and Ashhurst (1968) reported that in the basement membrane of muscle fibres of the stick insect, *Clitumnus extradentatus* (Cheleutoptera) banded collagen fibrils are to be found. Filaments are commonly found in the basement membrane of heart muscle fibres. For example, in the heart muscle fibres of the cockroach, *Blattella germanica* (Dictyoptera), the basement membrane is 130 mμ thick and filamentous, the filaments being arranged in a non-orderly feltwork (Edwards and Challice, 1960). In the region of the nerve-muscle synapses, the basement membranes of axon, glial cells, tracheoles and muscle fibre are fused, but basement membrane is apparently absent from the synaptic gap (Usherwood, 1967b). It is also interesting to note that basement membrane is either sparsely distributed within or absent altogether from much of the transverse tubular system (TTS) of locust leg muscle fibres (Cochrane *et al.*, unpublished).

The chemical composition of the basement membrane is not clearly understood. Ashhurst (1968) has suggested that acid muco-polysaccharides may be present in the basement membrane of flight muscle fibres of the water bug, *Lethocerus* (Hemiptera). If this is true, then the possibility that this structure is an ion exchange system must be seriously considered.

The plasma membrane of insect muscle fibres contains two functionally distinct components, an electrically inexcitable synaptic component and an electrically excitable (sometimes inexcitable) non-synaptic component (Usherwood, 1967b). Although these two components appear spatially separated and have different physico-chemical and pharmacological properties, it has not yet been possible to differentiate between them on a structural basis, although in the skeletal muscles of many insects the synaptic membrane can be identified in electron micrographs by the presence of underlying granules, the aposynaptic granules (Edwards *et al.*, 1958; Usherwood, 1967b) and in the sparseness or absence of basement membrane.

The plasma membrane of insect muscle fibres is invaginated at numerous points to form the TTS. As in crustacean muscle fibres (Peachey and Huxley, 1964; Brandt *et al.*, 1965) the TTS contains two morphologically distinct elements; sarcolemmal invaginations and tubules (Cochrane and Elder, 1967). The sarcolemmal invaginations are circular in cross section but the tubules are flattened (Hoyle, 1965). In the leg muscles of the locust, *Schistocerca gregaria* (Orthoptera), the larger elements, the sarcolemmal invaginations, enter at the Z-lines and send off branches vertically and horizontally. The other elements, the tubules, enter the fibres opposite the regions of overlap of the actin and myosin filaments but only penetrate the superficial part of the fibres. Basement membrane is restricted in the TTS to the main compartment of the sarcolemmal invaginations. The branches from the sarcolemmal invaginations and the tubules make contact with the SR to form the so-called "diadic junctions". The diad is a structurally differentiated region not unlike a "tight" junction (Hama, 1961) or ephapse (Bennett *et al.*, 1963). The diads are similar in appearance in both longitudinal and transverse sections. The membranes of the TTS and SR are separated by a space of about $6 \, m\mu$ to $14 \, m\mu$ and usually have the characteristic triple-layered "unit membrane" structure (Hoyle, 1965) although in the flight muscle of the beetle, *Tenebrio molitor* (Coleoptera) the membrane of the SR is apparently a simpler structure only $5 \, m\mu$ in width (Smith, 1961). In leg muscle fibres of *Schistocerca gregaria* (Hoyle, 1965; Usherwood *et al.*, to be published), a line of fine granules with regularly spaced thickenings are found between the membranes of the SR and TTS at the diadic junctions. Similar structures also occur at diads in cockroach leg muscle fibres (Hoyle, 1965).

B. ELECTRICAL PROPERTIES

Information on the electrical properties of insect muscle fibres is very sparse. In general terms the fibres can be considered as short leaky cables with infinite resistances at their ends. A summary of the electrical constants of a few insect muscles is given in Table I.

The cable constants were in all cases determined by the standard technique of applying small hyperpolarizing and depolarizing current pulses across the membrane of the muscle fibres and recording the resultant changes in membrane potential with either the same electrode, or another intracellular electrode, placed close by (ca 50 μ) the current electrode (Fatt and Katz, 1951). Most of the data presented in Table I was obtained using the cable equations of Hodgkin and Rushton (1946). Unfortunately these equations are applicable only when the length constant of the muscle fibre is small compared with the fibre length, so that the potential change resulting from the current pulse decays exponentially with distance from the current electrode. In many insect muscles the length constant of the fibre is long compared with the fibre length. Under these conditions the potential does not decay exponentially with distance from the current electrode and the "simple" cable equations do not apply. It is then necessary to use Weidmann's (1952) equations for a "cable" of short length. In most bioelectric tissues the specific resistance of the intracellular environment is usually 1.5 to 3 times higher than that of the external environment (Katz, 1966). If it is 3 times higher then the activities of some of the intracellular anions and cations may be lower than the concentrations of these ions, an important point to bear in mind when relating intracellular and extracellular ionic concentrations to membrane potentials.

C. PERMEABILITY

Compared with "stabilized" lipid bilayers (Hanai et al., 1965) muscle membranes have a low DC resistance which may indicate the presence of aqueous channels. The passage of water occurs equally fast through both lipid bilayers and muscle membranes and taking the evidence at face value it seems likely that water diffuses through the lipid phase.

A number of hypotheses have been invoked to account for the selective permeability of muscle membranes. The pore-solvent hypothesis as outlined by Shanes (1958) could be used as a starting point for those wishing to investigate this aspect further. Briefly, this proposes that the membrane contains aqueous channels, possibly with positive and/or negative fixed charge sites, which control the passage of ions. The channels are considered to show a dimensional variability consistent with the semi-rigid semi-fluid characteristics of cell membranes, and it is considered that large molecules would probably enter by other routes (e.g. by their ability to displace the membrane molecules laterally by intermolecular forces).

Transient changes in membrane permeability to ions may be due either to (a) changes in the diameter of channels through which unhydrated, incompletely hydrated, or fully hydrated ions move, or (b) changes in membrane-ion interactions (solvation). Since the membrane capacitance remains constant during excitation, the proportion of the surface area occupied by aqueous channels must be very small. Shanes (1958) has calculated for the squid axon membrane, a value of $2 \times 10^{-6}\%$ at rest and $8 \times 10^{-5}\%$ when excited, for the area occupied by aqueous channels. Alternatively the

P. N. R. USHERWOOD

Table I. Electrical consta

Insect species (order)	Muscle	Effective resistance [ohm] $(\frac{1}{2}\sqrt{r_m r_i})$	Specific resistance [ohm–cm^2] $\left(R_m = \frac{\lambda^2 2 R_i}{a}\right)$
Romalea microptera (A) (Orthoptera)	Skeletal (extensor tibiae)	$(1.9 \times 10^5)^{(1)}$	ca 3000
Schistocerca gregaria (A) (Orthoptera)	Skeletal (extensor tibia)	$8 \times 10^4 - 1.2 \times 10^6$	ca 8000
Locusta migratoria (A) (Orthoptera)	Skeletal (flight muscle)	—	ca 1500
Aeschna sp. (L) (Odonata)	Skeletal (leg muscles)	2.3×10^5	530
Galleria mellonella (L) (Lepidoptera)	Skeletal (ventral muscle)	$1.8 \times 10^4 - 1 \times 10^5$	ca 4000
Samia cecropia (A) (Lepidoptera)	Heart	3.8×10^6	1360
Periplaneta americana (A) (Dictyoptera)	Visceral (rectal muscle)	$5.5 \times 10^5 - 5 \times 10^6$	$2 \times 10^4 - 1.6 \times$
Antherea pernyi (L) (Lepidoptera)	Skeletal (leg muscle)	$2 \times 10^5 - 8 \times 10^5$	900–1600

(A) = adult   (L) = larva.
(1) Werman *et al.* (1961).

A value for R_i (the resistivity of the cell interior) of 250 Ω,
R_m is the resistance per unit area of membrane.
λ is the length constant of the muscle fibre.
τ_m is the membrane time constant.
a is the radius of the muscle fibre.

f insect muscle fibres

Membrane capacity $[\mu F/cm^2]$ $\left(C_m = \dfrac{\tau_m}{R_m}\right)$	Time constant [msec] (τ_m)	Length constant [mm] $(\lambda = (aR_m/2R_i)^{1/2})$	Fibre diameter $[\mu]$ (2a)	Reference
ca 2.2	6.6 (5.6–8.3)	1.7 (1.5–2.0)	ca 70–100	Cerf *et al.* (1959)
ca 2.0	16 (7–24)	2.8 (1.0–11.0)	ca 70–100	Usherwood (1962b)
ca 5.3	8.0	0.6–1.2	ca 70–100	Hagiwara and Watanabe (1954)
4.7	2.5	0.4 (0.2–0.6)	53	Malpus (1968)
ca 1.2	ca 6.0	0.56–1.47	flattened (ca 1 mm × 50μ)	Belton (1969)
—	—	ca 2.0	23	McCann (1965)
0.4–3.5	70	1.5–4.0	10	Belton and Brown (1969)
ca 4.0	8–12	0.6–0.8	ca 30–40	Belton (personal communication)

att and Katz, 1951) has been assumed for all muscles in this table.
$_m$ is the membrane resistance times unit length.
 is the specific resistance of the cell interior.
$_m$ is the membrane capacitance per unit area.

number of channels could be constant, activity merely leading to a marked decrease in resistance of pre-existing channels.

Until the chemical composition of the basement membrane around insect muscle fibres is known, it is difficult to ascertain the extent to which it serves as a barrier between these fibres and the haemolymph. It seems certain that it contains mucopolysaccharides (Ashhurst, 1968) but it is not known for certain whether these contain fixed charged sites. Rambourg and Leblond (1967) have suggested that the basement membrane around vertebrate cells contains both acid (negative charged sites) and neutral mucopolysaccharides, and that although it is readily permeable to water and inorganic ions, it can act as a sieve to large molecules. The presence of fixed charges (whether positive or negative) in the matrix could produce an ion exchange system with significant implications for the distribution of ions in the vicinity of the plasma membrane. Langley and Landon (1968) have demonstrated sulphated mucopolysaccharide surrounding the rat axon at the node of Ranvier and suggested that these fixed ionic sites might be involved in the maintenance of a high concentration of sodium outside the nodal membrane of the axon. Similarly Edwards *et al.* (1958) suggested that the basement membrane of the insect neurone serves to maintain a constant ionic environment at the plasma membrane. If this is the case then perhaps it would be useful to attempt to obtain a greater understanding of the role played by this membrane in ion distributions in insect muscles, especially in view of the possible absence of basement membrane in the fine branches of the TTS and in the synaptic clefts. As a starting point it might be worth while investigating the effects of removal of all or part of the basement membrane on the ionic properties of the muscle fibres. Further studies on the differences in ionic properties between areas of muscle membrane normally with and without basement membrane might also provide clues to the functional role of this structure. However, before embarking on these somewhat difficult lines of study it would perhaps be most useful to make further, more exhaustive, histochemical studies on the basement membrane itself.

III. THE EXTRACELLULAR AND INTRACELLULAR ENVIRONMENTS

Insect muscles are bathed in blood or haemolymph. The osmotic and ionic properties of insect haemolymph have been extensively studied and the vast range of published data on this subject has been reviewed on a number of occasions (e.g. Buck, 1953; Wyatt, 1961; Sutcliffe, 1963; Florkin and Jeuniaux, 1964; Stobbart and Shaw, 1964). Little purpose would be served in reviewing this information again in any great detail and therefore only a brief summary of the main osmotic and ionic characteristics of insect haemolymph will be given in this review. Table II summarizes some of the properties of haemolymph of representative members of the different insect orders.

In the lower insect orders, the ionic properties of the haemolymph are usually similar to those of the fluids which bathe vertebrate muscles, but in some phytophagous members of the primitive insect orders (e.g. Cheleutoptera) and in many members of the higher insects, the extracellular environment of the muscles is characterized by an ionic content which by vertebrate standards is very unusual. Florkin and Jeuniaux (1964) summarized the most striking peculiarities of the haemolymph of these higher insects as a tendency for replacement of inorganic osmotic effectors, usually sodium and chloride, by organic molecules, especially free amino acids and organic acids and the very special pattern of cationic composition, characterizing several orders. The cationic properties of the haemolymph of these insects are of especial interest since they may include one or more of the following: a very high potassium concentration; a low sodium concentration; a high magnesium concentration. It is true to say that vertebrate muscles could not function in an extracellular environment characterized by any one of these factors.

In the leg muscles of *Carausius morosus* the distribution of sodium in the extracellular and intracellular phases of the leg muscles is very unusual. Here the ratio $[Na_h^+]/[Na_i^+]$* is close to unity, due to the abnormally low extracellular sodium concentration. In this insect, magnesium assumes the role of the major extracellular cation. It is interesting to note that, although the sodium distribution is remarkably different from that seen in vertebrates, the potassium distribution is not unusual. This is not the case, however, in the larva of *Tenebrio molitor* (Belton and Grundfest, 1962) and in the moth, *Telea polyphemus* (Lepidoptera) (Huddart, 1966b). In the haemolymph of these insects the potassium concentration is unusually high, compared with the intracellular potassium concentration. The sodium distribution is also somewhat unusual, especially in the moth, where the intracellular sodium concentration greatly exceeds the

*$[Na_h^+]$ = concentration of sodium in haemolymph
$[Na_i^+]$ = concentration of sodium in muscle.

P. N. R. USHERWOOD

Table II. Osmotic and ionic properties of t▮

Order	Species	Δ°C	m.Osm	Inorganic constituen▮			
				Na⁺	K⁺	Ca⁺⁺	Mg⁺⁺
Ephemeroptera	*Ephemera danica* (L)	0.504	271	103	18	–	–
Odonata	*Aeschna grandis* (L)	0.735	395	145	9	7.5	7.5
Dictyoptera	*Periplaneta americana* (A)	0.897[1]	482	157	7.6	4.2	5.4
Isoptera	*Cryptotermes havilandi* (L)	0.730	392	103	28	8.6[2]	17.6[2]
Plecoptera	*Dinocras cephalotes* (L)	0.583	313	117	10	–	–
Cheleutoptera	*Carausius morosus* (A)	0.570[3]	306[3]	15	18	15	106
Orthoptera	*Locusta migratoria* (A)	0.841[6]	452[6]	60	12	17.2	24.8
Dermaptera	*Forficula auricularia* (A)	0.736	396	96	13	32.9[5]	–
Heteroptera	*Notonecta obliqua* (A)	0.756	406	155	21	15[7]	9.3[▮
Megaloptera	*Sialis lutaria* (A)	–	339	109	5	15	38
Neuroptera	*Myrmeleon formicarius* (L)	–	–	143.5	8.7	12.1	31.3
Trichoptera	*Phryganea* sp. (L)	0.455	245	69	7	14.4[8]	51[8]
Diptera	*Gastrophilus intestinalis* (L)	0.872	469	206	13	7	38
Lepidoptera	*Bombyx mori* (L)	0.480	258	13.9	38.7	18.1	50.3
Coleoptera	*Popillia japonica* (L)	1.03	554	20.2	9.5	7.9	19.4
Hymenoptera	*Apis mellifera* (L)	0.860	462	5.1	24	4.2	7.9

(1) Treherne (1961); (2) Clark (1958) (data for *Zootermopsis angusticollis*); (3) Ramsay (19▮ *Notonecta kirbyii*); (8) Duchateau *et al.* (1953); (9) Wyatt *et al.* (1956). (A) = adult; (L) = l▮

aemolymph of some representative insects

(mM)			m.Osm Inorganic cations	m.Osm Inorganic anions	Organic constituents (mM)		Reference
Cl^-	SO_4^{--}	HCO_3^-			Amino acids	Organic acids	
77	–	–	121	–	–	–	Sutcliffe (1962)
110	4	15	169	129	39	–	Sutcliffe (1962)
144	–	–	174.2	–	163	–	Van Asperen and Van Esch (1956)
82	–	–	–	–	–	–	Sutcliffe (1963)
111	4.1	–	–	–	39	–	Sutcliffe (1962)
101	16	5	–	35[5]	–	–	Wood (1957)
97.6	–	–	114	–	53[4]	–	Duchateau et al. (1953)
90	–	–	–	–	–	–	Sutcliffe (1963)
122	–	–	203.3	–	–	–	Sutcliffe (1962)
31	5[5]	15	167	51	85	–	Sutcliffe (1962)
–	–	–	196.5	–	–	–	Duchateau et al. (1953)
37	–	–	–	–	58	–	Sutcliffe (1962)
14.8	4	14.5	26.4	33.3	62	–	Levenbook (1950)
21	4.9	–	120	–	96[9]	–	Buck (1953)
19	4.9	–	80.9	–	170	–	Ludwig (1951)
33	10.3	–	84.5	–	207	–	Bishop et al. (1925)

Duchateau and Florkin (1958); (5) Clark and Craig (1953); (6) Sutcliffe (1963); (7) Clark (1958) (data for extracellular sodium concentration. Here, as in the stick insect, a divalent cation replaces sodium as the major cation in the extracellular phase.

Apart from inorganic ions, insects carry, dissolved in their haemolymph, substantial amounts of metabolites which other animals retain within their cells. Amino acids are present, often in very high concentrations, and on occasions probably account for the major ionic fraction of the haemolymph. A high aminoacidaemia is a characteristic of all insects but is more accentuated in endopterygotes. Considerable differences in the relative concentrations of individual amino acids occur in different species of insects and even in the same insect at different stages in its life cycle. Furthermore, there is evidence that the amino acid content normally fluctuates to compensate for changes in the inorganic ion content of the haemolymph. Other nitrogenous substances found in the extracellular fluid bathing insect muscles include ammonia, urea and uric acid, together with proteins. The quantity of protein has been estimated to be less than 5 gm/100 ml (Wigglesworth, 1965) being similar to that found in vertebrate plasma. Lipids, phospholipids and carbohydrates are also present, usually conjugated with proteins. The major blood carbohydrate is trehalose, which is often found in very high concentrations, for example in lepidopteran haemolymph.

The main organic acids found in insect haemolymph are the substrates of the enzymes of the tricarboxylic acid cycle. These acids (e.g. citrate and pyruvate) are more concentrated in larval haemolymph of endopterygote insects than in either adult haemolymph of these insects or the haemolymph of exopterygote insects.

In most animals the osmotic pressure of the extracellular fluid is maintained by regulating the univalent inorganic ion content (mainly chloride and sodium). In insects, however, inorganic cations and anions play a decreasing osmotic role as the evolutionary level of the insect increases (Sutcliffe, 1962, 1963). The osmotic pressure of insect haemolymph has been reported to range from values approximately equal to that of mammalian plasma, to twice that value. The osmotic pressure of the haemolymph can be strictly regulated, but not, presumably, by regulating the inorganic ion content (Florkin and Jeuniaux, 1964). Instead, changes in the amino acid content of the haemolymph occur, to compensate for fluctuations in inorganic ion concentrations. Ionic regulation does occur to some extent, however, since insects can maintain relatively constant levels of inorganic ions in their haemolymph in the face of substantial changes in the normal intake of these ions in the case of terrestrial insects and in the external environment in the case of aquatic insects (Stobbart and Shaw, 1964).

Although there is a wealth of information on the properties of the extracellular phase of insect muscle, information on the ionic and osmotic properties of the intracellular phase is at best very fragmentary. Analyses of the internal environment of insect muscle have been mainly restricted to the sodium and potassium content (Table III). In locusts, grasshoppers and cockroaches, the ionic content of the muscle fibres is essentially similar to that of vertebrate muscle (i.e. low sodium and chloride concentrations and a high potassium concentration).

Analysis of the ionic content of insect muscles, particularly the skeletal muscles, is a difficult and hazardous procedure. Since single insect muscle fibre preparations cannot yet be obtained routinely, muscles with only a few fibres are best for studies of this type, since problems of equilibration are then minimized. However, it is usually necessary to weigh the muscles in order to obtain an estimate of the ionic content and this is a very difficult exercise with small muscles such as the locust retractor unguis muscle (Usherwood, 1967b). Most studies of the ionic content of insect skeletal muscle have, therefore, been made using the large multifibre leg muscles of the larger insects, especially the extensor and flexor tibiae muscles of Orthoptera and Lepidoptera. Unfortunately, the problems of separating these large muscles from the exoskeleton and endoskeleton in order to extract them from the leg for weighing purposes, and of doing so without altering the ionic content of the fibres, have not been adequately faced as yet. Nor has the problem of equilibration of the extracellular and intracellular environments of these multifibre preparations following changes in the ionic properties of the extracellular phase.

Before leaving the general subject of the extracellular and intracellular environments of insect muscles, it would perhaps be advisable to mention the cellular content of insect haemolymph. Cells found in the blood of insects are usually called haemocytes and seem to have many important functions (e.g. clotting, phagocytosis etc.). Some insects, for example dipterous larvae (Arvy, 1952), are thought to have a cell-free haemolymph, while in other insects, the haemolymph contains up to 100,000 haemocytes/cm^3 (Tauber and Yeager, 1934, 1936). A detailed review on the properties and function of insect haemocytes has been made recently by Jones

Table III. Ionic characteristics

Insect species (order)	Phase	Inorganic constituents			
		Na^+	K^+	Ca^{++}	Mg^{++}
Periplaneta americana (A) (Dictyoptera)	Haemolymph	107.0	17.0	–	1.7
	Striated muscle	46.0	112.0	–	7.4
Periplaneta americana (A) (Dictyoptera)	Haemolymph	110.6	13.3	–	–
	Striated muscle	26.9	110.4	–	–
Locusta migratoria (A) (Orthoptera)	Haemolymph	103.0[2]	11.0[2]	–	–
	Striated muscle	18.8	124.0	–	–
Romalea microptera (A) (Orthoptera)	Haemolymph	64.0	18.0	–	–
	Striated muscle	44.0	128.0	–	–
Carausius morosus (A) (Cheleutoptera)	Haemolymph	15.0[3]	18.0[3]	–	–
	Striated muscle	12.9	103.3	–	–
Telea polyphemus (A) (Lepidoptera)	Haemolymph	2.5	54.1	–	72.0
	Striated muscle	17.6	77.3	–	–
Sphinx ligustri (A) (Lepidoptera)	Haemolymph	3.6	49.8	4.9	36.0
	Striated muscle	20.7	84.4	9.7	–
Telea polyphemus (A) (Lepidoptera)	Haemolymph	3.3	41.0	3.1	29.0
	Striated muscle	5.0	78.9	7.2	–
Bombyx mori (A) (Lepidoptera)	Haemolymph	9.0	41.3	7.7	42.3
	Striated muscle	12.4	97.7	16.7	–
Actias selene (A) (Lepidoptera)	Haemolymph	9.1	47.2	8.7	26.0
	Striated muscle	16.0	115.6	13.5	–
Samia cecropia (A) (Lepidoptera)	Haemolymph	2.5	54.0	–	–
	Heart muscle	20.0	88.0	–	–
Tenebrio molitor (L) (Coleoptera)	Haemolymph	75.5	36.5	–	–
	Striated muscle	18.0	72.0	–	–

E_{Na}, E_K, E_{Ca}, E_{Mg}, E_{Cl} = Equilibrium potentials for sodium
(1) Wood (1965); (2) Hoyle (1955); (3) Wood (1957).
  * Concentrations of ions in intracellular phases were originally expressed as m-moles/kgm tissue water.

of insect muscle systems

(mM)*	Predicted equilibrium potentials (mV)					Reference
Cl^-	E_{Na}	E_K	E_{Ca}	E_{Mg}	E_{Cl}	
$\overline{}$	+21.3	−47.5	−	+37.0	−	Tobias (1948)
95.7[1] 10.0[1]	+35.6	−53.3	−	−	−56.8[1]	Wood (1963)
93.9[1] 12.5[1]	+42.9	−61.0	−	−	−50.6[1]	Wood (1963)
$\overline{}$	+ 9.4	−49.4	−	−	−	Tobias (1948)
$\overline{}$	+ 4.6	−44.1	−	−	−	Wood (1963)
$\overline{}$	−49.3	− 9.0	−	−	−	Carrington and Tenney (1959)
63.1 14.5	−44.0	−13.3	−17.2	−	−36.3	Huddart (1966b)
67.5 15.0	−10.5	−16.5	−21.2	−	−37.9	Huddart (1966b)
68.2 13.3	− 8.0	−21.7	−19.1	−	−41.2	Huddart (1966b)
75.0 14.0	−14.2	−22.6	−11.1	−	−42.5	Huddart (1966b)
$\overline{}$	−52.5	−11.6	−	−	−	McCann (1965)
130.0–163.0 $\overline{}$	+36.1	−17.1	−	−	−	Belton and Grundfest (1962)

potassium, calcium, magnesium and chloride respectively.

(1964) and the reader is referred to this for further information. The only reason for mentioning here that these cells are present in insect blood is that estimations of extracellular ion concentrations using whole haemolymph, which includes the haemocytes, could lead to considerable errors in predicting muscle membrane potentials (Tobias, 1948). These errors would be even greater if the haemocytes played an ion regulatory role (Brady, 1967a, 1967b).

IV. THE RESTING MEMBRANE POTENTIAL

It is probably reasonable to assume that the bio-electric charac- teristics of most resting muscle cells result from the unequal distributions of the inorganic ions, sodium, potassium and chloride, on either side of the muscle membrane. The sign and size of the potential difference across the cell membrane is thought to be determined mainly by the relative permeabilities of these ions, although the activity of specific ionic pumps in the membrane, provided that they are electrogenic, may contribute to this potential difference. This is probably also the situation in muscles of many insects, for example those in which the extracellular environment of the muscles is not unlike that of verebrate muscles (Fig. 1). However, in insects which have, by vertebrate standards, unusually bizarre extracellular ionic environments, there is no reason to assume that sodium, potassium and chloride ions are involved in generating the resting potential across the muscle membrane. In these insects divalent cations may play a significant role in this respect and where the haemolymph contains high concentrations of organic compounds, it is possible that an unequal distribution and/or pumping of organic ions contributes to the potential difference recorded across the membrane of the muscle fibres. Of course to be effective in this respect, it is necessary that the organic particles must be charged and permeant.

Modern views of the origin of muscle resting potentials originated from a variety of sources which include Bernstein's ionic hypothesis (Bernstein, 1902, 1912); Gibbs-Donnan theory (Donnan, 1924) as presented by Boyle and Conway (1941); the Goldman constant field theory (Goldman, 1943) and numerous studies of active ion transport across the membrane of muscle cells (e.g. Keynes, 1954).

Bernstein considered that the resting membrane of bioelectric tissues was selectively permeable to potassium ions and impermeable to all other ions. Under these conditions the membrane potential

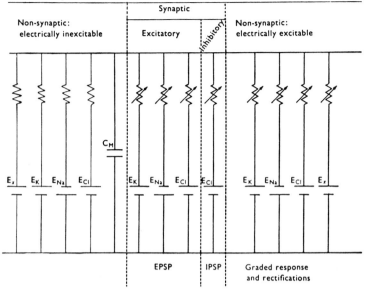

Fig. 1. Model of the electrical components of the membrane of the electrically-excitable muscle fibre of locusts and grasshoppers. C_M is the membrane capacity. E_K, E_{Cl}, E_{Na} are the potassium, chloride and sodium batteries (EMF = equilibrium potential). Ions other than sodium and potassium for the excitatory synaptic membrane component and sodium, potassium and chloride for the non-synaptic membrane component possibly carry some of the current during activity of these membrane components. The battery for these other ions, which could be either anions or cations, is denoted by E_x. (From Usherwood, 1968, modified after Grundfest, 1964.)

(E_M) would be equal to the potassium equilibrium potential (E_K), and could be calculated by knowing the external $[K_o^+]$ and internal $[K_i^+]$ potassium concentrations (provided that the activity) coefficient for this ion is the same on both sides of the membrane) and the absolute temperature of system (T), according to the Nernst relationship:

$$E_M \ (\equiv E_K) = \frac{RT}{nF} \ln\frac{[K_o^+]}{[K_i^+]} \quad \text{where } n = \text{valency (of potassium)}$$

$$F = \text{Faraday constant}$$
$$R = \text{gas constant.}$$

This membrane is to all intents a potassium electrode and the extracellular and intracellular environments constitute a potassium concentration cell.

Boyle and Conway (1941) later demonstrated that frog sartorius muscle fibres are permeable, when at rest, not only to potassium ions but also to chloride ions. Given this situation, it was suggested that the distribution of potassium and chloride ions was in accordance with a Donnan equilibrium, and that the resting potential difference across the cell membrane resulted from unequal distributions of potassium and chloride ions in the extracellular and intracellular phases of the muscle. Maintenance of this equilibrium was thought to depend on two factors: the impermeability of the membrane to sodium, the major extracellular cation, and the presence of impermeant anions in the intracellular phase. The intracellular impermeant anions are thought to consist mainly of organic phosphates and amino acids, with only 10% fixed to structural proteins. The presence of "fixed" negative charges in the intracellular phase and "fixed" positive charges (impermeant sodium ions) in the extracellular phase would constitute a "double" Donnan situation, and would influence the distribution of the permeant ions, potassium and chloride, so that at equilibrium, $[K_o^+] \times [Cl_o] = [K_i^+] \times [Cl_i^-]$ where $[Cl_o^-]$ and $[Cl_i^-]$ are the external and internal concentrations of chloride ions.

In a system containing three ionic species, one of which is impermeant, the number of osmotically active particles would be greater in the phase containing the impermeant ion, and the osmotic pressure of this phase would be greater than that of the phase containing only permeant ions. However, in frog muscle the presence of an effectively (see later) impermeant ion, sodium, in the extracellular phase "osmotically" balances the impermeant negative particles in the intracellular phase.

This idea of a simple "double" Donnan system required some modification with the discovery that the sodium permeability of the resting membrane, although much lower than the potassium and chloride permeabilities, was not zero. Therefore, in order to maintain the marked differences in intracellular sodium and potassium concentrations, either sodium ions and/or potassium ions must be actively transported across the membrane against their electro-chemical gradients. In actual fact, in frog skeletal muscle both these ions are transported across the membrane by a metabolically maintained "pump" whereas the distribution of chloride ions does not appear to involve an active transport process. If this is true, then the chloride distribution in this muscle is dependent on E_m and the concentration of impermeant anions in the intracellular phase. Even in the absence of the sodium-potassium pump, there should still be a potential difference across the membrane i.e.

$$E_M \ (\equiv E_{Cl}) = \frac{RT}{nF} \ln \frac{[Cl_i^-]}{[Cl_o^-]} \qquad \text{where } E_{Cl} = \text{the chloride equilibrium potential}$$

but the activities of sodium and potassium inside the muscle would now be equal. However, in this hypothetical situation the cell would have to contend with an osmotic gradient across the membrane.

In order to maintain the difference between $[K_o^+]$ and $[K_i^+]$, and $[Na_o^+]$ and $[Na_i^+]$ sodium and potassium ions are continuously transferred across the cell membrane against their electrochemical gradients. This is an active process involving the expenditure of energy (via ATP?) and transport of potassium into the cell appears to be coupled to transport of sodium out of the cell so that for every potassium ion entering by this metabolic "route" a sodium ion is extruded. A simple system involving active transport of only one of these cations would be sufficient to maintain the differences in the extracellular and intracellular distributions of these ions, but since it involves the unidirectional transport of a single charged particle the "pump" would be electrogenic. For example, extrusion of sodium from the cell without transfer of an equivalent charge in the opposite direction (i.e. into the cell) would set up a potential difference across the membrane (inside negative) which would sum algebraically with the diffusion potentials for potassium, chloride and sodium. Under these conditions the more permeant ions, potassium and chloride, would be distributed passively according to the EMF of the cell. A metabolic pump involving a one-for-one exchange of sodium and potassium ions, or a "pump" which extruded sodium with an anion, would be electrically neutral and the EMF of the cell would be determined by the distribution and relative permeabilities of all the permeant ions in the system. For this type of system

$$E_M = \frac{RT}{nF} \ln \frac{P_K [K_o^+] + P_{Na} [Na_o^+] + P_{Cl} [Cl_i^-]}{P_K [K_i^+] + P_{Na} [Na_i^+] + P_{Cl} [Cl_o^-]}$$

where P_{Na}, P_K and P_{Cl} are the relative permeabilities for sodium potassium and chloride respectively. In resting frog muscle $P_K > P_{Na}$ therefore E_M approaches E_K. Chloride ions are considered to be distributed passively according to E_M (i.e. $E_M \equiv E_{Cl}$). It is important to remember that the Goldman Equation used here refers to a "steady" state situation where the activities of sodium, potassium and chloride in the intracellular and extracellular phases are constant. Since both of the cations are permeant it is necessary to invoke "pumps" to maintain these constant ionic distributions (e.g. Hodgkin and Keynes, 1955).

In a few insect species, mainly belonging to the lower orders, there is evidence that the resting membranes of the skeletal (Usherwood, 1967a, 1967b) and visceral muscle fibres are potassium electrodes. Hagiwara and Watanabe (1953), Hoyle (1954) and Wood (1963) found that the membrane potential of locust muscle fibres varied with $[K_o^+]$ but their results indicated a considerable divergence from the slope of 58 mV to be expected for a pure potassium electrode when plotting E_M against $\log_{10}[K_o^+]$. However, Usherwood (1967b) demonstrated that locust skeletal muscle fibres are also permeable to chloride ions in the resting state (Fig. 2) and therefore the Nernst

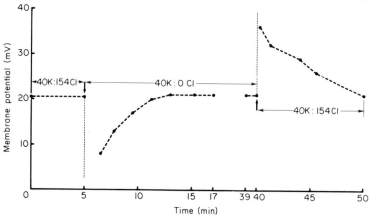

Fig. 2. Transient changes in membrane potential of a retractor unguis muscle fibre of *Romalea microptera* in response to variations in $[Cl_o^-]$. Muscle equilibrated in saline (40 mM potassium; 154 mM chloride). The "impermeant" anion propionate was substituted for chloride in the chloride-free saline. The similar time course for the depolarizing transient (chloride efflux) and hyperpolarizing transient (chloride influx) suggests that the time courses of the inward and outward chloride currents are more or less identical.

relationship for potassium is applicable to this system only when the intracellular and extracellular environments are in equilibrium (in steady state). If the system has not equilibrated then the relative conductances of the potassium and chloride ions must be known in order to predict E_M using the Nernst relationship. If sufficient time is allowed for complete equilibration then locust muscle fibres are found to behave like potassium electrodes (Usherwood, 1967b).

The permeability of locust skeletal muscle fibres to chloride ions was first demonstrated by Usherwood and Grundfest (1965). They recorded a large (ca 20 mV) transient depolarization of locust extensor tibiae muscle fibres when the chloride in the medium bathing these fibres was replaced by an "impermeant" anion. When chloride was reintroduced into the bathing medium a hyperpolarizing transient was observed (Fig. 2). Locust skeletal muscle fibres also swell when exposed to isosmotic saline containing a high concentration of potassium ions (Fig. 3). This swelling probably results from the osmotic influx of water accompanying the entry of KCl (Boyle

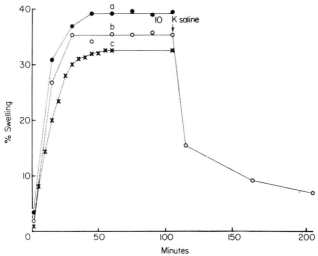

Fig. 3. Effect of changes of $[K_o^+]$ on volume of retractor unguis muscles of the locust, *Schistocerca gregaria*. At the start of the experiment three different muscles (a, b and c) soaked for 4 hours in low potassium saline (10 mM potassium 154 mM chloride) were perfused with high potassium saline (100 mM potassium, 154 mM chloride). The muscles increased in volume in the high potassium saline indicating that the membrane of the retractor unguis muscle is permeable to chloride and to potassium. After 100 min in high potassium saline muscle (b) was again exposed to the low potassium saline. Note rate of shrinkage in low potassium saline is much less than rate of swelling in high potassium saline, especially during later stages of shrinkage. When the KCl-swollen retractor unguis muscle is first exposed to low potassium saline it swells transiently (not shown on graph). This secondary, "anomalous" swelling has already been described for crayfish muscle fibres (Reuben *et al.*, 1964).

and Conway, 1941). Entry of KCl occurs if $[K_o^+]$ is increased whilst at the same time the product $[K_o^+] \times [Cl_o^-]$ is altered (increased). If $[Cl_o^-]$ is decreased when $[K_o^+]$ is increased, so that the product $[K_o^+] \times [Cl_o^-]$ remains constant, then the muscle does not normally swell (Usherwood, 1967a, 1967b, 1968). These results support the contention that locust skeletal muscle fibres are permeable in the resting state to potassium and chloride ions. Similar data have also been obtained from cockroach skeletal muscles (Usherwood, 1967b).

Locust and cockroach muscle fibres swell very quickly during KCl influx but when a swollen "KCl loaded" muscle is returned to a medium containing a low potassium concentration, KCl efflux is much slower (Fig. 3). It appears that, during KCl efflux, the permeability to either potassium or chloride or possibly both of these ions is reduced. In frog skeletal "fast" muscle fibres the permeability to potassium is reduced during KCl efflux (Hodgkin and Horowicz, 1957; Adrian, 1958, 1960). In both frog (Foulkes et al., 1965) and crustacean (Brandt et al., 1965) muscle, KCl efflux is associated with swelling of the transverse tubular system in the diadic (triadic in the case of frog muscle) regions and this is also true in locust skeletal muscle (Usherwood, 1967b; Cochrane and Elder, 1967). It must be remembered that when the external potassium or chloride concentra-

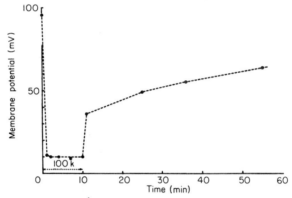

Fig. 4. Effect of changing $[K_O^+]$ at constant $[Cl_O^-]$ (154 mM) on the membrane potential of a grasshopper *(Romalea microptera)* retractor unguis muscle fibre equilibrated in low potassium saline (10 mM potassium; 154 mM chloride). The rate of depolarization when the fibre was exposed to high potassium saline (100 mM potassium) was much greater than the rate of repolarization when the muscle was returned to the low potassium saline, although the repolarization was characterized by an initial fast phase which took the membrane from about −10 mV to about −40 mV.

tion is changed, E_M will immediately take up a value which produces equal net movements of potassium and chloride ions in a system, such as locust skeletal muscle, where these are the major permeant ions. During KCl efflux from locust muscle fibres E_M initially falls very rapidly to a value approximately intermediate between the predicted values for E_K and E_{Cl} (Figs. 4 and 5). Therefore, at this stage it is reasonable to assume that P_K and P_{Cl} are of similar magnitude, since E_M will assume a value where the rate of potassium efflux and the rate of chloride efflux are equal. However, as $[Cl_i^-]$ falls, the rate of repolarization of the muscle fibres and rate of shrinkage of the muscle also falls, probably due to a reduction in

either potassium conductance or potassium and chloride conductance. Preliminary studies (Usherwood, to be published) of voltage/
current characteristics of locust retractor unguis muscle fibres in

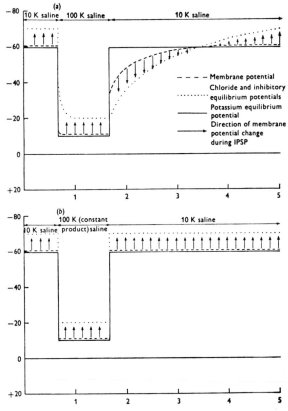

Fig. 5. Diagrammatic representation of effects on membrane potential (E_M), chloride
equilibrium potential (E_{Cl}) and potassium equilibrium potential (E_K) of a grasshopper or
locust skeletal muscle fibre of changes in the potassium concentration of the bathing
medium $[K_o^+]$. (a) When $[K_o^+]$ is changed whilst maintaining the chloride concentration of
the medium $[Cl_o^-]$ constant at 154 mEq/L, E_{Cl}, E_K and E_M change at different rates.
When $[K_o^+]$ is reduced under these conditions E_{Cl} becomes positive to E_M and the IPSP
(a chloride potential) becomes depolarizing. However, when $[K_o^+]$ is changed whilst
maintaining the produce $[K_o^+] \times [Cl_o^-]$ constant (b), E_K, E_{Cl} and E_M change to new values
simultaneously and the IPSP is hyperpolarizing for both increasing and decreasing $[K_o^+]$
(From Usherwood, 1968).

different potassium environments suggest that, for E_M values greater
(more negative) than -50 mV ($[K_o^+] < 12$ mEq/L), resistance to
inward (depolarizing) current is about 3 times greater than for

outward (hyperpolarizing) current (Fig. 6). This anomalous rectification could account for the asymmetrical time courses of the potential changes of locust muscle fibres to increases and decreases in $[K_o^+]$ (Fig. 4). For E_M values less than -50 mV the resistance to outward current is less than that to inward current. The threshold for

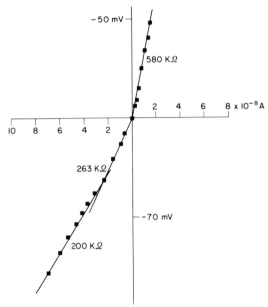

Fig. 6. Current-voltage relationship for a single muscle fibre of the metathoracic retractor unguis muscle of the grasshopper, *Romalea microptera*, illustrating anomalous rectification. Isolated muscle soaked for 4 hours in locust saline containing 10 mM potassium (Usherwood, 1968). Note greater conductance of muscle fibre during inward (hyperpolarizing) current than during outward (depolarizing) current. Current-voltage relationship was examined by passing depolarizing and hyperpolarizing current pulses (approximately 1 sec duration) across the membrane of the muscle fibre through an intracellular microelectrode. The resultant potential changes were recorded by a second intracellular microelectrode sited approximately 50 μ from the current electrode. When the muscle fibre was depolarized by more than 50 mV its resistance decreased markedly (delayed rectification). (Usherwood, unpublished observations.)

delayed rectification (potassium activation) is at an E_M of about -50 mV. Significantly the slope of the curve of the relationship between E_M and $\log_{10} [K_o^+]$ also changes at an E_M of -50 mV and for E_M values less than this the slope more nearly approaches 58 mV.

The sodium permeability of locust skeletal muscle fibres is assumed to be rather low, although not insignificant. The levelling out of the graph of E_M and $\log_{10} [K_o^+]$ for locust muscle fibres and for muscle fibres of other insects for low values of $[K_o^+]$ has been thought, in the past, to be due to a constant small permeability of the membrane to sodium ions (Wood, 1965). Usherwood (1967b) suggested that this may, in fact, be due to anomalous rectification and has found that if the system is given sufficient time to equilibrate then the slope of the relationship between E_M and $\log_{10}$ $[K_o^+]$ diverges only slightly from the 58 mV to be predicted for a potassium electrode membrane. However, in view of this slight divergence, it would perhaps be wise to use the Goldman Equation rather than the Nernst relationship to predict E_M for low values of $[K_o^+]$ (Hodgkin and Katz, 1949).

Since the Goldman Equation represents a steady state situation where the concentrations of sodium, potassium and chloride ions in the extracellular phases are constant, it must be assumed that for even very low values of P_{Na} some metabolic process is required to maintain these concentrations (Fig. 7). There is no evidence that a coupled sodium-potassium pump such as is found in frog skeletal muscle fibres exists in insect skeletal muscle fibres. The occurrence of hyperpolarizing inhibitory postsynaptic potentials (IPSP's) in locust and grasshopper muscles indicates that some active outward transport of chloride is required if P_{Cl} is high in the resting muscle fibre, since production of the IPSP involves chloride activation (Usherwood and Grundfest, 1965). Hyperpolarizing IPSP's involving chloride activation also occur in cockroach skeletal muscle fibres (Usherwood and Grundfest, 1965) and these fibres are also permeable to chloride when at rest (Usherwood, 1967b). Therefore, a coupled outward sodium-chloride "pump" has certain attractions for these muscles, since it could be electrically neutral.

If this sodium-chloride pump was only loosely coupled, so that the chloride transport (extrusion) was unaffected by reduction of $[Na_o^+]$, then it would be possible to explain the fall in E_M of cockroach, locust and stick insect muscle fibres when $[Na_o^+]$ is reduced (Wood, 1957, 1963), since the chloride pump would then become electrogenic.

Huddart and Wood (1966) found that E_M in the skeletal muscle fibres of *Periplaneta americana* begins to decline almost immediately after treatment with DNP, a metabolic inhibitor (Fig. 8A). The curve relating E_M to time of exposure to DNP is complex but indicates, nevertheless, the marked dependency of E_M on metabolic processes. It does not indicate, however, that there is an electrogenic pump in this muscle, since DNP does not cause an immediate significant fall in

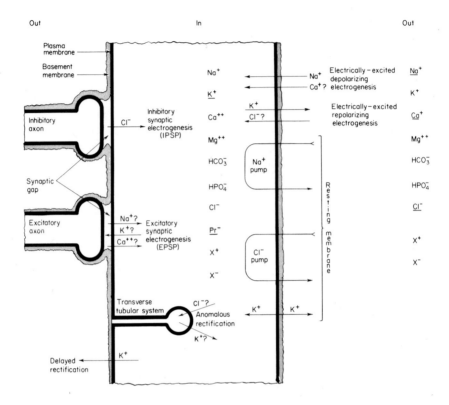

Fig. 7. Hypothetical schematic representation of the ionic properties of a locust skeletal muscle fibre. *At rest:* The muscle membrane is a good potassium electrode ($E_M = E_K$). Potassium ions are considered to be distributed passively since there is as yet no evidence for a potassium pump in this system. E_{Cl} is more negative than E_M and E_{Na} is more positive than E_M. Sodium and chloride ions are actively distributed by metabolic pumps which are coupled and electrically neutral. Negatively charged impermeant protein particles (Pr^-) make up the bulk of the intracellular anion concentration. X^- and X^+ represents amino acids and organic acids etc. The distribution of these has not yet been determined, nor has their role in determining the electrochemical characteristics of the muscle system. *During activity:* There are four types of electrically-excited responses. Two of these, the depolarizing electrogenesis and the repolarizing electrogenesis are the components of the graded, spike-like responses of locust muscle fibres. The other two electrically-excited responses are represented by anomalous rectification and delayed rectification. Although these two "rectification" responses are spatially separated in this scheme there is as yet no evidence that this occurs in insect muscle. The chemically-evoked responses of the electrically-inexcitable synaptic membrane are either excitatory (depolarizating) or inhibitory (hyperpolarizing).

The role of the basement membrane in determining ionic distributions and/or membrane properties is not known although the absence of this structure at the postsynaptic membranes and in the finer elements of the transverse tubular system may account in part for the highly specific permeability of these regions.

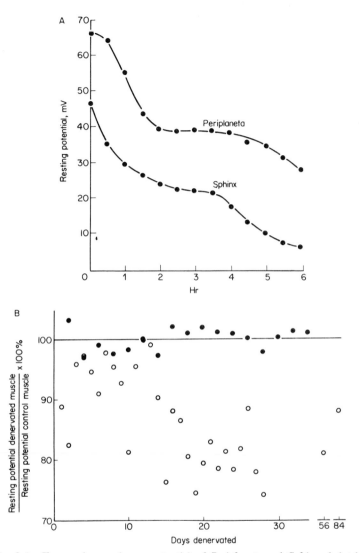

Fig. 8 A. Changes in membrane potential of *Periplaneta* and *Sphinx* skeletal muscle fibres under the influence of DNP (0.5 mM). (From Huddart and Wood, 1966.)

B. Effect of denervation on the resting membrane potential of locust metathoracic extensor tibiae muscle fibres. The major fall in membrane potential of the denervated fibres coincides with degeneration of the terminals of the innervating motor axons, and failure of transmission at the nerve-muscle synapses. The contralateral (right) metathoracic extensor tibiae muscles of the experimental animals were used as controls. Comparison of right and left metathoracic extensor tibiae muscles of normal animals shown as filled circles. Resting membrane potentials of denervated fibres shown as open circles. (After Usherwood, 1963b.)

E_M; either sodium is exchanged for another external cation (e.g. potassium) or it is extruded with an anion (e.g. chloride). The decline in E_M of locust skeletal muscle fibres following denervation (Usherwood, 1963b, 1963c) could be due to failure of a metabolic "pump" or "pumps" to maintain the normal distributions of ions across the muscle fibre membrane (Fig. 8B).

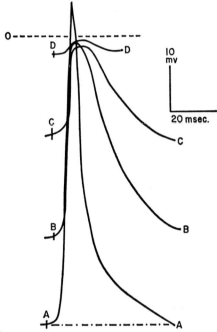

Fig. 9. The depolarizing effect of carbon dioxide on the skeletal flight muscle fibres of a wasp. Original resting level indicated by dash-dotted line: O, zero potential; A, normal neurally-evoked response (EPSP plus electrically-excited depolarization); B–D, response at different times after treatment with 100% CO_2. Stimulation was at 1/sec. Depolarization was reversible and complete within about 3 sec. (From McCann and Boettiger, 1961.)

It is generally assumed that the membrane potentials recorded from bioelectric tissues are the result of a Donnan distribution of permeant ions and possibly also partly due to the activity of electrogenic metabolic "pumps". In insects the studies by McCann and Boettiger (1961) on the flight muscles of *Sarcophaga bullata* (Diptera), the flesh fly, and *Vespula diabolique* (Hymenoptera), a wasp, are potentially of considerable significance in this respect. They found that carbon dioxide almost completely depolarized these muscles (Fig. 9). It is possible, of course, that by exposing these muscles to a stream of 100% carbon dioxide McCann and Boettiger inadvertently altered the pH of the bathing medium (haemolymph). Usherwood (unpublished observation) has found that pH has a marked effect on E_M of locust leg muscle fibres. However, it is possible that E_M of the fly and wasp flight muscle fibres studied by McCann and Boettiger (1961) is completely metabolically maintained and that carbon dioxide blocks the source of energy required to maintain this metabolism. There seems little doubt that these muscle fibres deserve further examination.

One way of examining the permeabilities of muscle fibres to potassium, sodium and chloride ions is to measure both E_M and the internal and external concentrations of these ions using muscles exposed to saline containing different amounts of sodium, potassium and chloride. Wood (1963) measured $[K_o^+]$ and $[K_i^+]$ of leg muscles of *Locusta, Periplaneta* and *Carausius* after soaking them for 4--6 hours in saline of inorganic composition approximating to that of haemolymph. In each case E_M was close to the value predicted by the Nernst relationship for $[K_i^+]$ and $[K_o^+]$. When, however, Wood (1965) varied $[Cl_o^-]$ of locust and cockroach muscle fibres he found that the chloride equilibrium potential (E_{Cl}), calculated after analyses of the chloride content of the muscles, differed significantly from E_M except for $[Cl_o^-]$ of 10 mM. Wood (1965) suggested that one possible explanation for this observation is that, in the resting muscle fibres of these insects, the permeability to chloride ions is low. However, Usherwood (1967a, 1967b) has clearly shown that this is not the case. The fact that Wood (1965) did not get a fall in $[K_i^+]$ when he reduced $[Cl_o^-]$ is not particularly surprising since, even if the membrane is permeable to both potassium and chloride, the total KCl efflux during reduction in $[Cl_o^-]$ will be small and will be accompanied by water efflux. Therefore, in view of the normally high internal potassium ion concentration no significant change in $[K_i^+]$ would be expected, nor would any permanent change in E_M be expected following changes in $[Cl_o^-]$, only transient depolarizations (reduction in $[Cl_o^-]$) or hyperpolarizations (increase in $[Cl_o^-]$). Wood (1965) found that when $[Cl_o^-]$ is reduced from 100 mM to 0 mM, $[Cl_i^-]$ did not change significantly. However, it is perhaps pertinent to point out that there was a significant fall in $[Cl_i^-]$ when $[Cl_i^-]$ was reduced from 150 mM to 100 mM. Wood (1965) used the large and complex extensor and flexor tibiae muscles of the locust, *Locusta migratoria* (Orthoptera) and *Periplaneta americana* and it is possible that his results reflect to some extent the difficulties of equilibrating these muscles with saline. It is possible that chloride efflux from skeletal muscle fibres in these insects may occur across the membrane of the TTS (Brandt *et al.,* 1965; Cochrane and Elder, 1967; Usherwood, 1967b). Swelling of the TTS has been observed in cockroach and locust muscle fibres (Cochrane and Elder, 1967) during KCl efflux and it is possible that this is due to accumulation of chloride in this extracellular compartment. If this is true, then Wood's (1965) analyses of $[Cl_i^-]$ could include extracellular chloride within the swollen TTS. On the basis of his results, Wood (1965) concluded that a Donnan equilibrium involving potassium and chloride ions may not occur in locust and cockroach muscles. It is probably correct to say that a true Donnan distribution for potassium and chloride does not occur in these muscles since there is evidence that the chloride distribution is actively maintained, although the chloride pump is not able to maintain $[Cl_i^-]$ constant in the face of changes in the product $[K_o^!] \times [Cl_o^-]$. If one accepts his ionic distribution data then the discrepancies between Wood's predicted values for E_M, and the actual values he recorded with microelectrodes could be due to his assumption that the concentrations of chloride ions and potassium ions inside the muscle fibres are equal to the activities of these ions. Reuben *et al.* (1964) have shown that some crustacean muscle fibres are not characterized by a Donnan distribution for potassium and chloride ions, although they are undoubtedly permeable to these ions when at rest. On the other hand Hays *et al.* (1968) have recently found that muscle fibres of *Callinectes* (the blue crab) appear to obey the requirements for a Donnan distribution of ions on the basis of electrophysiological and osmotic data, but their data on ionic concentrations had to be corrected for non-solvent water, bound potassium and bound chloride fractions before they also agreed with the Donnan distribution hypothesis. The problem of bound ion fractions in insect muscle has not yet been studied but in view of the findings of Hays *et al.* (1968) it might be valuable to re-investigate the Donnan relationships of locust and cockroach muscle fibres although I am convinced that the large leg muscles used by Wood (1963, 1965) are most unsuitable for this type of study.

The effect of sodium ions on the resting membrane potential of muscle fibres of lower insects is not yet clear. Wood (1957) found that changes in $[Na_o^+]$ affected the magnitude of E_M of skeletal muscle fibres of *Carausius morosus;* E_M increased when $[Na_o^+]$ was increased. $[Na_o^+]$ is normally low in this insect although slightly greater than $[Na_i^+]$ and it is possible that when $[Na_o^+]$ is increased it enhances the activity of a sodium "pump". This could convert a previously neutral "pump" into an electrogenic pump. Alternatively, an existing electrogenic pump may, under these conditions, make a greater contribution to E_M. The fall in E_M of grasshopper leg muscle fibres (Werman et al., 1961) (where $[Na_o^+]$ is much greater than $[Na_i^+]$), when sodium is replaced by barium in concentrations >10 mM could also be explained by changes in the relative activities of sodium, potassium and chloride pumps (or a pump for other inorganic and organic ions for that matter).

In higher insects it is unlikely that in the resting muscle E_M is determined by the distributions of potassium and chloride. In many of these insects $[Na_o^+]/[Na_i^+]$ is often less than unity, $[Cl_o^+]$ is often very low and the amino acid content of the extracellular phase (haemolymph) is often high (Tables II and III). $[K_o^+]$ can also be very high, yet this does not appear to markedly affect the resting membrane potential. For example, Belton and Grundfest (1962) found that muscle fibres of the mealworm, Tenebrio molitor, which are normally surrounded by haemolymph containing a high concentration of potassium ions are insensitive to changes in $[K_o^+]$ between 1 mEq/L and 120 mEq/L and even when these fibres are exposed to saline containing 240–320 mEq/L potassium there is little change in E_M for 30 mins. Measurement of $[K_o^+]$ and $[K_i^+]$ for mealworm muscle fibres in haemolymph gave a value for E_K of -17 mV i.e. much lower than the potential difference recorded from the muscle fibres (ca -50 to -60 mV) (Table III). However, E_M falls with a slope of 30 mV per decade increase in $[K_o^+]$ when the anion species in the extracellular phase is sulphate, phosphate or isethionate, rather than chloride (Grundfest, personal communication). It seems probable that the membrane of mealworm muscle fibres behaves as a mixed electrode system for several ions, anions as well as cations, but not for divalent cations (Grundfest, personal communication). E_M increases with increasing concentrations of chloride, acetate, propionate and nitrate, but decreases with increasing concentrations of isethionate and sulphate. Possibly isethionate and sulphate are impermeant anions, although in view of the complex electrochemistry of these muscle fibres it would be advisable to study volume and/or weight changes during transfer of the muscle from chloride to isethionate or sulphate to test this possibility more vigorously. E_M of mealworm muscle fibres also depends to some extent on $[Na_o^+]$ (i.e. E_M changes by 25 mV for a decade change in $[Na_o^+]$).

Apart from the mealworm, the only other insects which have been studied in any great detail in terms of muscle electrochemistry are members of the Lepidoptera. Moth skeletal muscles appear to be insensitive to changes in $[K_o^+]$ over a wide range of potassium concentrations (Huddart, 1966c) (Fig. 10A-B). There is also apparently little correlation between the observed resting E_M and the predicted E_K based on measurements of intracellular and extra-cellular concentrations of potassium (Table III). Huddart (1967) also found that E_{Cl} is closer to E_M (Fig. 10C). Apparent differences in E_K and E_{Cl} must, however, be viewed with some caution until it can be clearly demonstrated that measurements of internal and external potassium and chloride concentrations are for an equilibrated (steady state) system (Usherwood, 1967b). The fact that different authors have described widely differing values for resting membrane poten-tials of lepidopteran muscle fibres suggests that this equilibrium state has not always been achieved at the time of measurement (Belton, 1958; Van der Kloot, 1963; Carrington and Tenney, 1959; Duchateau *et al.*, 1953; Huddart, 1966a). The problems of equili-bration are poignantly illustrated by an observation made by Belton (personal communication). He found that when the small retractor unguis muscle from a moth leg is immersed in *distilled water* it still contracts for up to 3 hours as a result of the continued activity of the central fibres of the muscle. This muscle contains only a few fibres. Is there any value, therefore, in continuing with studies of the effects of ions on insect muscle using large muscles such as the flexor and extensor tibiae muscles?

Although there is some doubt about the accuracy of estimates of $[K_o^+]$ and $[K_i^+]$ for mealworm (Belton, and Grundfest, 1962) and moth (Carrington and Tenney, 1959; Huddart, 1966b) muscles, for reasons outlined above, E_K for these muscles does appear to be very low (ca -20 mV) and far removed from E_M (ca -60 mV). Either the resting membrane is not at all permeable to potassium ions or the potassium distribution is actively maintained against a low resting leakage of potassium from the muscles. Huddart and Wood (1966) believe that the potassium gradient across some lepidopteran muscle fibres is at least partly actively maintained. For example, the slow decline in E_M of the moth, *Sphinx* (Lepidoptera) skeletal muscle fibres during treatment with DNP suggests that the ionic distri-butions across the membrane are actively maintained (Huddart and Wood, 1966), although the absence of any marked initial fall in E_M during DNP treatment apparently rules out any direct contribution to E_M by an electrogenic pump.

In summary, it appears that the skeletal muscles of the higher

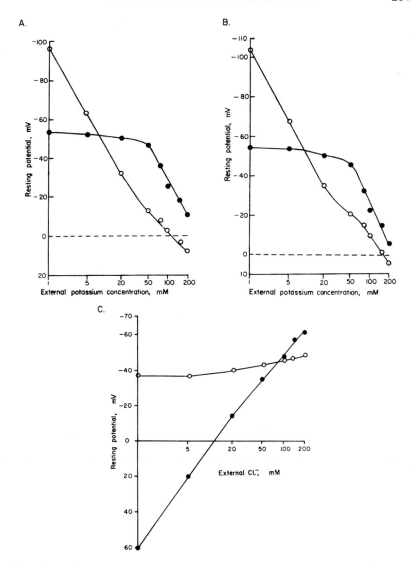

Fig. 10 (A–B). Relationship between resting membrane potential (full circles) and E_K (hollow circles) (predicted from data on the concentrations of potassium in the intracellular and extracellular phases) and $[K_o^+]$ for skeletal muscle fibres of (a) *Sphinx ligustri* and (b) *Actias selene*. (From Huddart, 1966d.)

(C). Relationship between resting membrane potential (hollow circles) and E_{Cl} (filled circles) of *Sphinx ligustri* skeletal muscle fibres. E_{Cl} predicted from data on the concentrations of chloride in the intracellular and extracellular phases. (From Huddart, 1967.)

insects studied so far are at rest multi-ionic electrodes, although the exact nature of the role of various ions in determining E_M are not yet known. In muscles of these insects as well as in insect muscles with simpler electrode properties, considerable care must be exercised when investigating the effects on E_M of a single ion species. Unless the roles of all the other ions involved in producing and maintaining E_M are considered at *all times,* completely erroneous conclusions may be drawn from data obtained from these muscles.

The resting potential of moth heart muscle fibres appears, at first sight, to be determined mainly by the unequal distribution of sodium ions in the intracellular and extracellular phases (i.e. $E_M = -50$ to -55 mV whereas $E_{Na} = -52$ mV and $E_K = -11$ mV (McCann, 1965) (Table III). However, the close correspondence between E_M and E_{Na} may be purely coincidental since when $[Na_o^+]$ is reduced to zero, E_M does not apparently change.

The haemolymph of many of the higher insects stands in dramatic contrast to the blood and plasma of most lower insects and vertebrates not only in its unusual cation composition but also in its anion constitutents. Chloride is often not the predominant anion and the extracellular phase may contain such divalent anions as sulphate and also a number of organic anions (Table II). McCann (1964) studied the effects of several anions on the E_M of moth heart muscle fibres and concluded that the order of permeabilities of these anions is acetate>nitrate>bromide>chloride. However, until more information is available on the roles of other ions, cations and anions, in the production of resting E_M of moth heart muscle, it would be wise to view these conclusions with some reserve. If the resting membrane is a pure anion electrode, then substitution of chloride by other anions will transiently hyperpolarize or depolarize the muscle fibres according to whether the ions are more or less permeant than chloride. However, if the resting membrane is a multi-electrode system, then changes in E_M following anion substitution would be more difficult to interpret. There is also a danger here of considering the distribution of anions and cations as passive and ignoring the possibility of the presence of electrically neutral or electrogenic pumps. In fact, McCann and Wira (cited by McCann, 1967) have evidence for a potassium pump in moth heart muscle, and McCann (1967) has remarked on the possible presence of other pumps (i.e. for anions as well as for cations). Moth heart muscle is sensitive to DNP (10^{-4}M) and sodium azide, falling (to a constant level) by about 60% in 10 mins (McCann, 1967). The fall in E_M with DNP is accompanied by efflux of potassium and it has been suggested that this is good evidence for a potassium pump. Possibly this pump is electrogenic. Alternatively, it could be electrically neutral if potassium uptake was coupled with uptake of an anion. A coupled sodium-potassium pump is unlikely in this system since the sodium and potassium concentration gradients are in the same direction (McCann, 1965). The effects of metabolic inhibitors on the E_M of moth heart muscle certainly need further examination, together with more extensive analysis of the distribution of organic and inorganic anions and cations in the extracellular and intracellular environments of the muscle fibres.

Unfortunately, there is little information on the electrochemistry of insect gut muscle and reproductive muscle, and it is only very recently that the first recordings from insect gut muscle fibres have been obtained using intracellular electrodes (Belton and Brown, 1968). These authors recorded from the muscle fibres of the cockroach rectum and obtained values for resting membrane potential averaging about -30 mV, although they did obtain values as high as -53 mV. Since the skeletal muscle fibres of the cockroach are good potassium electrodes (Usherwood, 1967b) and in saline containing 2.5 mM potassium such as that used by Belton and Brown (1968), give values for E_M in excess of -70 mV, it is possible that the visceral fibres have different electrode characteristics.

Until more information is available on the ionic and osmotic properties of insect muscle, it would perhaps be advisable to keep extrapolations from vertebrate muscle to insect muscles to a minimum. Assuming that the Goldman Constant Field Theory applies to insect as well as to vertebrate systems, the complexity which is found to be a property of the electrochemistry of muscle in at least some of the higher insects would have to be expressed by the expanded equation:

$$E_M = \frac{RT}{F} \ln \frac{P_{Na} [Na_o^+] + P_K [K_o^+] + P_{Cl} [Cl_i^-] + P_{Ca} [Ca_o^{++}]^{1/2} +}{P_{Na} [Na_i^+] + P_K [K_i^+] + P_{Cl} [Cl_o^-] + P_{Ca} [Ca_i^{++}]^{1/2} +}$$

$$\frac{P_{Mg} [Mg_o^{++}]^{1/2} + P_I [I_o^+]^{1/2} + P_I [I_i^-]^{1/2} + P_O [O_o^+]^{1/2} + P_O [O_i^-]^{1/2}}{P_{Mg} [Mg_i^{++}]^{1/2} + P_I [I_i^+]^{1/2} + P_I [I_o^-]^{1/2} + P_O [O_i^+]^{1/2} + P_O [O_o^-]^{1/2}}$$

where $[I_i^+]$, $[I_o^+]$, $[I_o^-]$, $[I_i^-]$ = internal and external concentrations of unidentified inorganic cations and anions; similarly [O] for organic ion concentration; n = valency.

Although as a first approximation it may seem reasonable to

emphasize the importance of the relative concentrations of permeant cations and anions in the extracellular and intracellular phases of insect muscle, it should be borne in mind that the strict thermo-dynamic derivation of the Nernst and Goldman type would require the use of ion activities rather than ionic concentrations. Where extracellular or, for that matter, intracellular ionic concentrations do not seem to support a Gibbs-Donnan distribution and/or E_M cannot be predicted on the basis of the relative permeabilities and concentrations of the permeant ions in the system, it is possible that the activities of at least some of these permeant ionic species are lower than their concentrations. Ion binding could account for this. Ion binding may occur even in the extracellular phases of some muscles. For example, Carrington and Tenney (1959) concluded that about 10% calcium and 20% magnesium in the haemolymph of the moth, *Telea polyphemus* (Lepidoptera) is bound to organic particles, whereas Weevers (1966) found that nearer 60% of the calcium and magnesium in the haemolymph of this insect is bound, and also obtained indirect evidence for potassium binding.

One major problem in this respect, which in the past has received little attention, is the role of the haemocytes in "controlling" the free ion content of insect haemolymph. Tobias (1948) found that cockroach serum contains less sodium, potassium and magnesium than whole haemolymph. Brady (1967a, 1967b, 1968) believes that in the cockroach, *Periplaneta americana,* the haemocytes contribute very significantly to the level of potassium in the blood, but since this haemocyte-potassium fraction is compartmentalized it would not affect the resting E_M of the skeletal and visceral muscle fibres. Perhaps haemocytes sequester other ions as well as potassium. This is one aspect of the problem of insect electrochemistry that definitely deserves further very careful consideration.

V. SYNAPTIC MEMBRANES

The input component in the membrane of bioelectric tissue is, as a rule, electrically inexcitable (Grundfest, 1956, 1957). In muscle tissue, the input component is the postsynaptic membrane i.e. the area of muscle membrane which responds to transmitter released from the terminals of the innervating axons. Bennett (1964) aptly defined this electrically inexcitable membrane component as mem-brane which "behaves as a voltage source of constant EMF with conductance which is an increasing function of the transmitter concentration and independent of the applied field across the membrane". In other words, the reaction between the excitatory (and inhibitory) transmitter and the postsynaptic membrane receptors has no regenerative link (Katz, 1962).

At synapses between muscle fibres and excitatory neurones it has been invariably found that presynaptic impulses evoke a localized depolarization of the postsynaptic membrane, the excitatory post-synaptic potential (EPSP) (Eccles, 1964). This is a transient fall in the potential difference across the postsynaptic membrane and consists of a relatively rapid rising phase followed by a slower, quasi-exponential, decay phase. During activation, the excitatory postsynaptic membrane is depolarized as a result of an inward flow of current in that part of the membrane acted on by the transmitter, with a corresponding outflow of ionic current across the remainder of the muscle membrane. In frog skeletal muscle fibres, the excitatory synaptic current was at one time assumed to be due to a general increase of ionic permeability of the postsynaptic membrane (Fatt and Katz, 1951), but is now thought to be due to activation of sodium and potassium ions only (Takeuchi and Takeuchi, 1960). In these muscle fibres the postsynaptic potential (always excitatory) reaches a peak in 1 msec and lasts about 20 msec. The transmitter (acetylcholine) action is most intense during the first 2–3 msec and lasts a total of no more than 5 msec. There is a linear relationship between E_M and EPSP and between E_M and the current which flows during the EPSP (Takeuchi and Takeuchi, 1960). These relationships seemingly establish that the synaptic current is due to ions moving down their electrochemical gradient rather than to changes in activity of a metabolic pump (Eccles, 1964). When E_M is reduced towards zero the magnitude of the EPSP falls and the response reverses to a hyperpolarizing potential at an E_M between -10 mV and -20 mV. The reversal potential, or more correctly equilibrium potential (E_{EPSP}), is the value for E_M at which no net current flows during activation of the postsynaptic membrane, but net ionic fluxes may occur if more than one ionic species is involved. It may be a little unwise at this stage to rule out completely the possibility that the synaptic response of muscle results from action of the transmitter on a metabolic pump. For example, Kernan (1967) suggests that the end-plate potential of rat skeletal muscles might be generated by the action of the transmitter, acetylcholine, on a potassium transport system thus activating an electrogenic pump which depolarizes the muscle membrane. In support of his suggestion, Kernan (1967) points out that in the skeletal muscle fibres of the rat the spontaneous miniature endplate potentials are suppressed by hypoxia (Hubbard and Løyning, 1966) and that removal of extracellular sodium reduces the amplitude of the end-plate potential and also inhibits the potassium pump in the muscle.

Electrogenesis at inhibitory synapses in vertebrates, and in invertebrates, involves an increase in the conductance of the postsynaptic membrane component for potassium and/or chloride ions. In insect muscle (Usherwood and Grundfest, 1965), and for that matter crustacean muscle (Boistel and Fatt, 1958; Grundfest, 1959), the inhibitory synaptic current is carried by chloride ions alone. Like the EPSP, the inhibitory postsynaptic potential (IPSP), when it occurs, has a fast rising phase and a slow falling phase. In some systems, for example crustacean muscle fibres, activation of the inhibitory postsynaptic membrane by the transmitter released from the endings of the inhibitory neurone does not always involve a change in E_M. Here the equilibrium potential for the IPSP is equal to E_M and the inhibitory action is mediated through the increased conductance of the muscle fibre which results from interaction between the inhibitory transmitter and the inhibitory postsynaptic membrane. At most inhibitory synapses, however, inhibitory electrogenesis involves the production of a hyperpolarizing IPSP with an equilibrium potential about 10 mV more negative than E_M.

A. PROPERTIES OF EXCITATORY SYNAPTIC MEMBRANES

It has been customary, in the past, to divide the excitatory synaptic responses of insect skeletal muscle fibres into "fast" EPSP's and "slow" EPSP's. The adjectives "fast" and "slow" refer to the relative speed of the mechanical responses accompanying the synaptic potentials and not to the velocity of conduction along the motor axon or to the characteristics of the EPSP's (Hoyle, 1952, 1957). Unfortunately these terms are often now used in a somewhat different context to distinguish the electrical responses of insect muscle fibres on the basis of their relative magnitudes, large responses being termed "fast"; small responses "slow". Since the excitatory responses of insect muscle fibres come in all shapes and sizes, perhaps it would be sensible, if somewhat pedantic, to drop the terms "fast" and "slow" and describe the responses as EPSP's or EPSP's plus electrically-excited responses, as the case may be, at the same time giving details of their main distinguishing characteristics.

Del Castillo *et al.* (1953) were the first to study the excitatory synaptic responses of insect skeletal muscle fibres. Using locust leg muscle they found that if E_M is raised (made more negative) or lowered, by passing polarizing or depolarizing current across the membrane of a muscle fibre, the amplitude of the EPSP is correspondingly raised or lowered and that there is a simple linear relationship between EPSP amplitude and E_M (Fig. 11A and B). The EPSP's do not exhibit refractoriness.

The ionic basis for excitatory synaptic electrogenesis in insects has not been studied as yet in any great detail. In species where the ionic content of the haemolymph is not unlike that of the extracellular environment of vertebrate skeletal muscle it is possible that sodium and potassium ions carry the current during activation of the excitatory postsynaptic membrane (Fig. 7). Kusano and Grundfest (1967) suggested that sodium and, to some extent, divalent cations, contribute to synaptic depolarizing electrogenesis in mealworm muscle fibres where the haemolymph has very different ionic properties from vertebrate plasma. Here E_{EPSP} is close to zero and potassium and chloride ions do not contribute to the synaptic current. Kusano and Grundfest (1967) based their suggestion of sodium activation during the EPSP on the assumption that lithium cannot substitute for sodium during the EPSP. Nerve-muscle transmission in the mealworm is blocked when all the sodium in the extracellular environment is replaced by lithium. However, it is possible that failure of nerve-muscle transmission was due to the action of lithium on the presynaptic, rather than postsynaptic, membrane. Indeed, in the locust retractor unguis muscle lithium appears to block transmission at the excitatory synapses presynaptically, since in saline containing lithium instead of sodium the muscle responds to glutamate treatment but not to neural stimulation (Machili and Usherwood, unpublished).

In many insects, particularly those containing only low concentrations of sodium in the haemolymph, removal of sodium from the extracellular environment of the muscle has little, if any, effect on the EPSP. This is true for flight muscle fibres of the moth, *Bombyx mori* (Lepidoptera) (Huddart, 1966a) and for muscle fibres of the moth, *Antherea pernyi* (Lepidoptera) (Belton, personal communication). Sodium certainly appears to carry at least part of the current during excitatory synaptic electrogenesis in the leg muscles of the grasshopper. For example, Werman *et al.* (1961) found that for muscles of this insect when all the sodium in the bathing medium

A

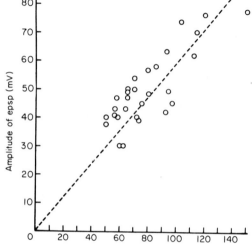

B

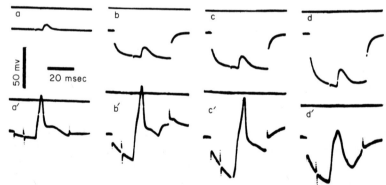

Fig. 11 A. Relationship between amplitude of EPSP of locust *(Schistocerca gregaria)* flexor tibiae muscle fibres and membrane potential of muscle fibres. According to these data the equilibrium potential estimated from the reversal potential of the EPSP is close to 0 mV. (From del Castillo *et al.*, 1953.)

B. Different effects of hyperpolarization on EPSP's and electrically-excited responses of grasshopper *(Romalea microptera)* extensor tibiae muscle fibre. a–d: the EPSP was increased in magnitude by hyperpolarizing the muscle fibre with current from an intracellular electrode. a'–d': in another fibre a large EPSP plus electrically-excited response was separated by the same procedure. Hyperpolarization (b', c') at first augmented both the EPSP and the EPSP remained unaffected. (From Cerf *et al.*, 1959.)

was replaced with sucrose there was no longer any evidence of a response to neural stimulation. Divalent cations, such as barium and strontium could substitute for sodium during the EPSP to a limited extent, although the synaptic response was depressed during prolonged treatment with strontium. In skeletal muscles of the stick insect, *Carausius morosus,* Wood (1957) found that the rate of rise of the EPSP was affected by the concentration of sodium in the bathing medium, the rate of rise of the synaptic response increasing with increases in $[Na_o^+]$. Increases in $[Na_o^+]$ also produce an increase in magnitude of the EPSP. This is not an unexpected result if sodium normally contributes to the EPSP.

The relative roles of divalent cations and organic ions in electrogenesis at insect excitatory nerve-muscle synapses are not very clear. In vertebrate muscle the limits of variation in the extracellular concentrations of calcium needed to counteract the blocking action of magnesium are about half that of the corresponding blocking magnesium concentration, up to the limit 20 mM magnesium: 10 mM calcium (Hoyle, 1955). The main synaptic action of magnesium and calcium ions in insects which have haemolymph containing low concentrations of these ions is on the release of transmitter from presynaptic nerve terminals, calcium enhancing, and magnesium depressing the release of this chemical (Usherwood, 1963a). However, magnesium also has an effect on the muscle membrane in these insects. In concentrations greater than 10 mM it reduces the amplitude of the miniature EPSP's of locust muscle fibres, which result from the release of the transmitter from the excitatory nerve terminals (Usherwood, 1961, 1963a; Usherwood and Machili, 1968; Usherwood *et al.,* 1968). It is possible that magnesium decreases the effective resistance of these fibres. The fact that glutamate potentials can still be evoked at synapses on locust muscle fibres when nerve-muscle transmission has been blocked by removal of extracellular calcium suggests that this ion contributes very little to the synaptic current.

Neuromuscular synaptic transmission in insects which normally contain high concentrations of magnesium ions in their haemolymph has been studied by a number of workers. Belton (personal communication) found that in skeletal muscle fibres of *Antherea pernyi* 0–70 mM of magnesium in the bathing medium had no effect on the EPSP, although there was a reversible reduction in the magnitude of the EPSP after 30 min in saline containing 70 mM magnesium. The EPSP's of muscle fibres of *Carausius morosus* were also unaffected when exposed to high concentrations of magnesium ions although there was a reduction in the EPSP when $[Mg_o^{++}]$ was increased to 200 mM (Wood, 1957). Weevers (1966) found that the size of the EPSP of *Antherea pernyi* skeletal muscle fibres was exaggerated when the fibres were exposed to a high concentration of magnesium. The haemolymph of this insect normally contains a high level of magnesium, although Weevers (1966) suggests that at least 60% of this is bound.

In caterpillar skeletal muscle fibres the EPSP is reduced as $[Ca_o^{++}]$ is increased, the rate of reduction of the EPSP being directly proportional to the frequency of stimulation of the innervating axons (Belton, personal communication). Calcium ions affect the size and rate of rise of the EPSP in leg muscles of the stick insect, presumably by influencing the release of transmitter from the nerve terminals (Wood, 1957). The EPSP in this insect is greatly depressed in zero calcium saline but is not abolished, even after prolonged exposure of the muscle to saline containing no calcium (Wood, 1957).

It is obvious that studies of the ionic basis for depolarizing synaptic electrogenesis in insects are still in a very elementary stage and that the results obtained so far are very fragmentary. What is now required is a detailed examination of single synapses using either voltage clamp techniques or extracellular recording techniques for studying the synaptic currents. In this manner it may be possible to differentiate clearly between the effects of ions on synaptic and non-synaptic membranes of insect muscle fibres.

Studies of the pharmacological properties of the excitatory postsynaptic membrane of insect muscle fibres are worthy of mention, if only because of the recent important advances made in this field. Evidence derived from a number of independent studies strongly indicates that the excitatory nerve-muscle synapse in insect skeletal muscle is not cholinergic (e.g. Iyotami and Karneshina, 1958; Harlow, 1958; Hamori, 1961; Hill and Usherwood, 1961), but that transmission at this synapse involves a dicarboxylic amino acid (Usherwood and Grundfest, 1965; Kerkut et al., 1965; Kerkut and Walker, 1966; Usherwood and Machili, 1966; McCann and Reece, 1967; Usherwood et al., 1968a; Usherwood et al., 1968b; Usherwood and Machili, 1968). Usherwood and Machili (1968) studied the pharmacological properties of excitatory nerve-muscle synapses on the retractor unguis muscle of the locust and found that L-glutamate was the most active dicarboxylic amino acid. When L-glutamate is applied iontophoretically to the synaptic sites the postsynaptic

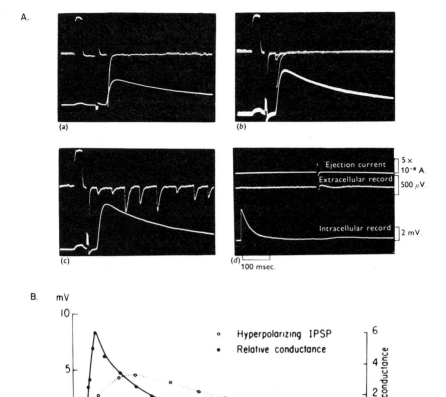

Fig. 12 A. Intracellular (lower traces) and extracellular (upper traces) EPSP's recorded from a locust retractor unguis muscle fibre (resting membrane potential, 60 mV). The extracellular EPSP's were recorded from glutamate-sensitive sites on the muscle fibre. Preparation perfused with saline containing 40 mM Mg^{++} (substituted for Na^+) to reduce height of EPSP's so that the electrically excitable membrane of the muscle fibre was not excited and muscle contractions were abolished. Note brief duration of extracellular EPSP (synaptic current) compared with intracellular response. Note also variable height of EPSP (b). In (c) the neurally evoked extracellular EPSP is accompanied by a burst of spontaneous miniature EPSP's. The extracellular and intracellular responses to a small quantity of iontophoretically-applied glutamate recorded from the same synapse as (c) are illustrated in (d). No synaptic current (extracellular EPSP) accompanied the intracellular EPSP indicating that transmission at this synapse had failed on this occasion. Calibration pulse at the beginning of each record (a–c) was 1 mV; 5 msec. (From Usherwood and Machili, 1968.)

B. Comparison of the time course of the IPSP of grasshopper and locust skeletal muscle fibres (open circles, *ordinate* on left) with the time course of the increase in membrane conductance. Membrane conductance changes are expressed in arbitrary units (*ordinate* on right) as the decrease in amplitude of a testing pulse which was caused by an intracellularly-applied brief current of constant amplitude delivered to the muscle fibre at different times during the IPSP. (From Usherwood and Grundfest, 1965.)

membrane responds gradedly, the magnitude of the depolarization being directly proportional to the concentration of glutamate ejected from the glutamate source (Usherwood and Machili, 1968; Usherwood *et al.*, 1968b; Beránek and Miller, 1968a, 1968b). Significantly, the equilibrium potential from the glutamate response is close to that for the EPSP (10–25 mV inside negative) (Beránek and Miller, 1968a, 1968b). Usherwood *et al.* (1968b) and Usherwood and Machili (1968) also showed that the synaptic current, recorded by applying an extracellular microelectrode close to an excitatory synapse, flows mainly during the rising phase of the EPSP suggesting that the duration of transmitter action on the postsynaptic membrane is brief, at locust excitatory synapses (Figs 12 A and 13).

Agents that block synaptic electrogenesis do not usually affect either the conductance of the resting muscle membrane nor the resting potential. The only substances known to block transmission at insect excitatory synapses are tryptamine and its analogues (Hill and Usherwood, 1961; Usherwood, 1963a; Aidley, 1967).

EPSP's have already been recorded from skeletal muscle fibres of many different insect species. Recently T. Miller (personal communication) has recorded EPSP's from cockroach heart muscle fibres, although here the excitatory postsynaptic membrane does not appear to be sensitive to glutamate. EPSP's have also been recorded recently from cockroach rectal muscle fibres following stimulation of the proctodeal nerve (Belton and Brown, 1969). Very little is known of the pharmacology of insect visceral muscles (Davey, 1964), although

---

Fig. 13 A. Diagrammatic representation of the voltage changes during the neurally-evoked and electrically-excited responses of insect skeletal muscle fibres. The EPSP is usually of shorter duration than the IPSP (see text). The spike-like graded electrically-excited response consists of a depolarizing component and a repolarizing component. The repolarizing component may contain either a pronounced negative after potential (hyperpolarization) or a pronounced positive after potential (depolarization). A second spike-like response may appear on a positive after potential.

B. Diagrammatic representation of the conductance changes underlying the above responses. The duration of the conductance increased during the IPSP is greater than that during the EPSP suggesting prolonged transmitter action at the inhibitory synapses, although part of this conductance increase could be due to anomalous rectification in either potassium or chloride channels. Chloride carries the inhibitory synaptic current but the ions involved in excitatory synaptic electrogenesis are not yet known. In some insects the depolarizing component of the electrically-excited response is thought to involve an increase in sodium conductance whilst the repolarizing component is thought to involve an increase in potassium conductance. The graded nature of this electrically-excited response is thought to be due to an early increase in potassium conductance. Early inactivation of potassium will produce a positive after potential; late inactivation of potassium will produce a negative after potential if the resting membrane potential is less negative than E_K.

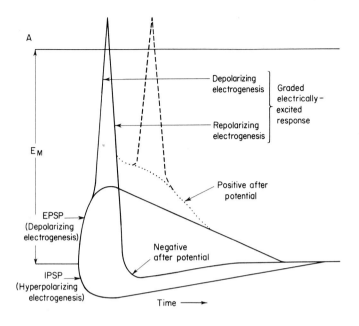

A

Depolarizing
electrogenesis

Repolarizing
electrogenesis

} Graded
electrically –
excited
response

E_M

Positive after
potential

EPSP
(Depolarizing
electrogenesis)

Negative
after potential

IPSP
(Hyperpolarizing
electrogenesis)

Time ⟶

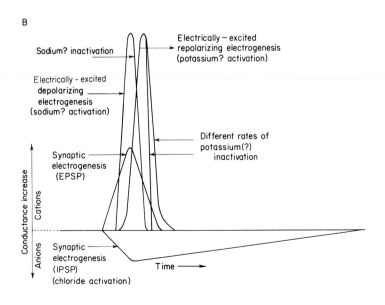

B

Sodium? inactivation

Electrically – excited
repolarizing electrogenesis
(potassium? activation)

Electrically – excited
depolarizing
electrogenesis
(sodium? activation)

Different rates of
potassium(?)
inactivation

Synaptic
electrogenesis
(EPSP)

Conductance increase

Cations

Anions

Synaptic
electrogenesis
(IPSP)
(chloride activation)

Time ⟶

the hind gut muscle of *Schistocerca gregaria* is sensitive to γ-aminobutyric acid (Dowson and and Usherwood, to be published), the probable transmitter at inhibitory synapses in insect skeletal muscles (Usherwood and Grundfest, 1965).

B. PROPERTIES OF INHIBITORY SYNAPTIC MEMBRANES

Inhibitory postsynaptic potentials (IPSP's) have so far been recorded from muscles of locusts, grasshoppers, cockroaches (Usherwood and Grundfest, 1964, 1965; Hoyle, 1966a, 1966b), bees and beetles (Ikeda and Boettiger, 1965a, 1965b) and there now seems little doubt that peripheral inhibitory systems are probably very widespread throughout the insect group. In the locust meta-thoracic extensor tibiae muscle (Usherwood and Grundfest, 1965) and metathoracic coxal adductor muscle (Hoyle, 1966a, 1966b; Usherwood, 1968) the IPSP's are exclusively hyperpolarizing potentials provided the fibres have completely equilibrated (steady state) with the extracellular environment of the muscle (Fig. 5). The amplitude of the IPSP is normally correlated partly with the magnitude of the resting potential although there are obvious differences in the amount of transmitter released from endings in different fibres and in the "sensitivity" of the inhibitory post-synaptic membrane on these fibres. For example, Usherwood and Grundfest (1965) found that in some fibres the IPSP was sometimes a maximal event, whilst in other fibres the IPSP was sub-maximal and repetitive stimulation of the inhibitory neurone led to facilitation of the IPSP's.

In locust and grasshopper skeletal muscles the duration of the IPSP ranges between 50–300 msec. The IPSP is usually of greater duration than the EPSP (15–150 msec) (Usherwood and Grundfest, 1965). Although the major increase in conductance during the IPSP is brief, there is a small maintained increase in conductance lasting about as long as the hyperpolarization itself (Fig. 12B and 13). This suggests that the interaction of the inhibitory transmitter with the postsynaptic receptors is a prolonged event, possibly due to the absence of an enzyme for degrading the transmitter. However it is possible that some of the increase in membrane conductance which occurs during the IPSP is due to anomalous rectification resulting from hyperpolarization of the muscle fibre membrane although, this would only be true when the resting E_M is >-50 mV (Fig. 6).

The reversal (equilibrium) potential for the IPSP (E_{IPSP}) in locust and grasshopper muscle fibres is about -70 mV in saline containing 10 mM potassium (Usherwood, 1968). With higher and lower concentrations of potassium than this in the external environment E_{IPSP} is less negative and more negative accordingly (Usherwood, 1968).

The synaptic current during the IPSP of locust and grasshopper leg muscle fibres is mainly carried by chloride ions (Usherwood and

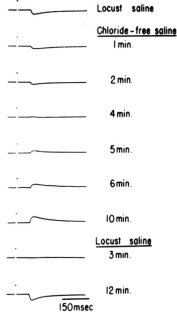

Fig. 14. Reversal of IPSP of grasshopper extensor tibiae muscle fibre in chloride-free saline. A transient depolarization which occurred on substituting propionate-saline for the standard chloride saline is not shown in the records. However, the IPSP became smaller despite this depolarization and then reversed to become a depolarizing response. When the muscle was returned to the chloride saline the IPSP reversed to become a hyperpolarizing response once again. (From Usherwood and Grundfest, 1965.)

Grundfest, 1965; Usherwood, 1968). When the chloride in the medium bathing is replaced by the supposedly impermeant anion, propionate, the IPSP reverses to a depolarizing response and E_{IPSP} becomes positive (Fig. 14).

Usherwood and Grundfest (1965) stated that E_{IPSP} is not changed significantly by substituting potassium for sodium in the extracellular environment of a grasshopper or locust muscle. Usherwood (1968) has found, however, that E_{IPSP} does change if $[K_o^+]$ is increased, provided sufficient time is allowed for re-equilibration of the system (i.e. influx of KCl and water into the muscle fibres). Entry of additional chloride significantly increases $[Cl_i^-]$. As a result, E_{Cl} and therefore E_{IPSP} becomes less negative but since E_M has also fallen the IPSP is still a hyperpolarizing event. When a muscle loaded with KCl is returned to a medium containing a low concentration of potassium the IPSP's transiently become depolarizing responses since E_{Cl} becomes transiently less negative than E_M (Usherwood, 1968) (Fig. 15).

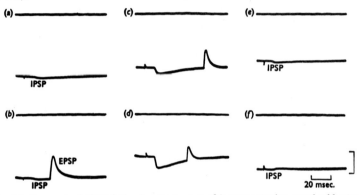

Fig. 15. Reversal of IPSP following treatment of locust anterior coxal adductor muscle with high potassium saline (40 mM potassium, 154 mM chloride). Reflexly evoked IPSP's (a–f) and spontaneous EPSP's (b–d) recorded intracellularly from a single superficial anterior coxal adductor muscle fibre. Muscle soaked for 1 hour in low potassium saline (10 mM potassium, 154 mM chloride). Responses at end of equilibration period shown in a–b. The IPSP initially increased in magnitude when high potassium saline was substituted for the low potassium saline (c–d) but the EPSP became smaller as the fibre depolarized. When the low potassium saline was reintroduced the IPSP became smaller as the muscle repolarized (e) and then reversed to a depolarization (f). Note different durations of EPSP and IPSP. Time and voltage calibrations were the same for all records. (From Usherwood, 1968.)

The pharmacological properties of the insect inhibitory postsynaptic membrane have been the subject of preliminary studies by Usherwood and Grundfest (1964, 1965). They found that γ-aminobutyric acid (GABA) activates the inhibitory postsynaptic membrane and that picrotoxin blocks activation of this membrane both by GABA and the inhibitory transmitter. GABA appears to have no effect on the non-synaptic membrane and therefore is inactive in systems which lack an inhibitory innervation. It is interesting to note that the chemically sensitive membranes, i.e. the postsynaptic membranes of both excitatory (Fig. 16A) and inhibitory (Fig. 16B–E) synapses, in insects appear to be restricted to the subsynaptic areas of the muscle fibre membrane. However, Usherwood (1969) has recently demonstrated that when locust muscle fibres are denervated the entire muscle membrane becomes sensitive to L-glutamate.

VI. ELECTRICALLY EXCITABLE MEMBRANES

In the discussion of the structural properties of insect muscle membranes, it was pointed out that most insect muscle fibres have two functionally distinct membrane components. One of these components is electrically excitable (i.e. its electrogenic reactions are in response to electrical stimuli). According to the broad definition of the term electrical excitability adopted by Grundfest (1966) "any activation or inactivation process leading to a change in membrane permeability that results from an electrical stimulus constitutes an electrically-excited response so that, for example, such phenomena as delayed rectification and hyperpolarizing responses must be included here". Presumably the phenomenon of anomalous rectification must also be included.

The general view on the nature of the electrical responses of bioelectric cells is that these responses are due primarily to changes in permeability which may be highly specific for an individual ion species or alternatively involve more than one ion. Provided that the ion or ions involved are not in electrochemical equilibrium when the membrane is activated these changes in permeability produce diffusion potentials and there will be a net movement of these ions down their electrochemical gradient. If the ions involved are in electrochemical equilibrium when the membrane is activated, then an increase in permeability to these ions may not involve any net current flow and, if this is the case, then no potential change will be observed.

Although this reviewer adopts the generally accepted view that potential changes in bioelectric cells usually result from changes in membrane permeability, it must always be borne in mind that this is not the only possible explanation. Tasaki, for example, has challenged this viewpoint on a number of occasions and the reader is referred to his many reviews (e.g. Tasaki and Singer, 1965, 1968) and excellent book (Tasaki, 1968) outlining his objections to the "equivalent circuit" or "ionic" theory.

In the majority of vertebrate muscle cells the most obvious expression of electrically-excited activity is the all-or-none action

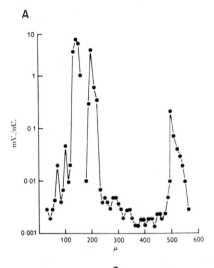

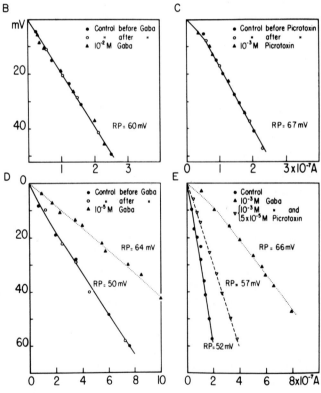

potential. Shanes (1958) defines the action potential as "the sequence of more or less rapid changes in potential difference ... that begins with the familiar spike best known for its all-or-none properties under normal conditions". Voltage clamp studies have demonstrated that in the squid giant axon the action potential results from a short duration surge of inward current carried by sodium ions, followed by a more prolonged but smaller outward current carried by potassium ions (Hodgkin and Huxley, 1952). In other words, according to the "ionic theory", the all-or-none action potential results from two activation processes in which the permeability of the membrane, first to sodium ions and then to potassium ions, is greatly increased. During sodium activation the membrane is depolarized and E_M approaches the sodium equilibrium potential (E_{Na}) so that at the peak of the action potential:

$$E_M \ (\equiv E_{Na}) = \frac{RT}{nF} \ln \frac{[Na_o^+]}{[Na_i^+]}.$$

Repolarization results partly from the delayed increase in P_K and partly from a fall in P_{Na} (sodium inactivation). With more or less simultaneous potassium activation and sodium inactivation the membrane approaches a pure potassium electrode (i.e. E_M approaches E_K) so that

$$E_M \ (\equiv E_K) = \frac{RT}{nF} \ln \frac{[K_o^+]}{[K_i^+]}.$$

If the resting membrane is also a pure potassium electrode then the repolarization due to potassium activation returns E_M to its resting condition but if, as in some cells, E_M is less negative than E_K then a

Fig. 16 (A). Distribution of glutamate-sensitivity along part of an anterior coxal adductor muscle fibre of the locust *(Schistocerca gregaria)*. The glutamate was ejected electrophoretically from a micropipette and the sensitivity of the muscle fibre membrane was measured as the depolarization in mV per nC of charge used to displace the ionized amino acid from the pipette. The pipette was moved in $10\ \mu$ steps along the fibre surface with the exception of a small portion of the membrane covered with a tracheole. (From Berànek and Miller, 1968b.)

(B-E). Different effects of GABA and picrotoxin on effective membrane resistance and membrane potential of four adjacent fibres of the metathoracic extensor tibiae muscle of the grasshopper *(Romalea microptera)*. *Ordinates* represent the hyperpolarization caused in each fibre by intracellularly-applied currents. *Abscissae* represent membrane potential with resting membrane potentials at origins of graphs. (b--c) Two fibres which were not innervated by an inhibitory axon. Neither the membrane potential nor effective resistance of these fibres was affected by GABA (b) or picrotoxin (c) applied in high concentration. (d--e) Two fibres innervated by an inhibitory axon. The two fibres were hyperpolarized by applying GABA (d) and their resistance decreased. These effects were reversed either by washing out the GABA (d) or by applying picrotoxin in the presence of GABA (c). (From Usherwood and Grundfest, 1965.)

phase of hyperpolarization (negative after-potential) may occur before the resting state is re-established. The process of sodium inactivation together with the process of potassium activation contribute to the refractoriness of the membrane (i.e. its inability to respond for some time after a spike) (Hodgkin and Huxley, 1952).

The increase in P_{Na} and P_K during the action potential and the temporary depolarization of the muscle fibre and reversal in polarity of E_M would be expected to lead to an influx of sodium ions and an efflux of potassium ions. Although the amounts of these ions transferred across the membrane during an impulse are small (ca 10^{-12} moles/cm), some mechanism must be present to remove the sodium ions from the cell and to reincorporate the lost potassium ions into the intracellular phase so as to recharge the sodium and potassium batteries. If this does not occur then the capacity of the cell to give a normal size action potential (ca 140 mV) will diminish with each successive impulse. The sodium and potassium batteries are in fact recharged by the sodium-potassium pump.

It is appropriate at this point to insert a slight note of caution. The ideas on the ionic basis for generation of the action potential briefly outlined above have mainly evolved from studies of squid giant axons. Electrophysiological studies of many varieties of cells have disclosed electrogenic phenomena which differ considerably from the spike electrogenesis of the squid giant axon (e.g. Grundfest, 1966). Although most, if not all, of these electrically-excited responses can be explained using equivalent circuit theory, it would be unwise to treat all bioelectric tissues as if they were squid giant axons. This is especially true in the insects where not only are the ionic environments of the muscle cells considerably different from those of the squid axon but the electrical responses of insect muscle membranes are often qualitatively different in many respects.

The familiar all-or-none action potential or muscle spike does not normally occur in insect muscle fibres, at least in those studied so far. Here the electrically excitable membrane of the muscle fibre is gradedly responsive (Figs 11B and 13). As stated earlier, most electrophysiologists use the term *action potential* to describe a transient *all-or-none* depolarization of a bioelectric membrane which propagates *without decrement* from its point of origin. According to this definition, therefore, the *graded* electrically-excited response characteristic of insect muscle fibres is not an action potential since its amplitude is dependent upon the magnitude of the electrical stimulus and it propagates with decrement. The graded electrically-excited response does however have some features in common with the action potential, for example it shows refractoriness (Cerf *et al.*, 1959), and it can be converted into an all-or-none action potential under certain conditions (Werman *et al.*, 1961; Usherwood, 1962b; Hoyle, 1962). A possible explanation for graded responsiveness of the electrically excitable membrane of the insect muscle fibre was given by Werman *et al.* (1961). They suggested that if the resting P_K was sufficiently high or if the increase in P_K triggered by electrical stimulation occurred early enough, the depolarizing regenerative response initiated by inward movement of sodium or for that matter any other ion, would produce only a highly damped response (Fig. 13). Two important features of this graded response are worth mentioning, namely that the depolarizing component is regenerative, and, perhaps what is more important, the peak depolarization of the graded response does not necessarily represent an equilibrium potential for the ion or ions responsible for the depolarizing electrogenesis. This last point cannot be emphasized too strongly since confusion of graded with all-or-none responses has in the past led to a number of possibly misleading conclusions about insect muscle systems. It is undoubtedly true that graded electrically-excited responses sometimes overshoot the zero membrane potential but for this to occur a very large EPSP must first be generated by the synaptic membrane. Multiterminal or distributed innervation as it is sometimes called, ensures that the entire membrane of an electrically excitable insect muscle fibre is depolarized to approximately the same extent during neural stimulation since the electrically excitable membrane is "stimulated" by the current generated at all the synaptic sites. All-or-none responses are therefore not essential for this system.

Del Castillo *et al.* (1953) introduced the term "active membrane response" to differentiate the graded electrically-excited response of locust and cockroach skeletal muscle fibres from the EPSP. Although this term has since been widely used by many insect electro-physiologists, including the author of this review, it is really quite misleading since EPSP's, IPSP's, delayed rectifications and anomalous rectifications are equally responses of active (activated) membranes.

The depolarizing response of the electrically-excited component of the insect muscle membrane is graded in amplitude according to the stimulus intensity, and propagates with decrement from the site of stimulation. This electrically-excited response is subject to refractori-ness, the absolute refractory period for the skeletal fibres of the grasshopper, *Romalea microptera* (Orthoptera), lasting at least 7 msec and relative refractoriness over 25 msec (Cerf *et al.,* 1959). Stimulation of an electrically excitable muscle fibre with a long intracellularly applied depolarizing current pulse evokes a train of responses which increase in frequency as the stimulus is increased (Fig. 17).

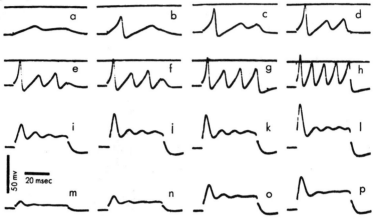

Fig. 17. Direct excitation of a grasshopper *(Romalea microptera)* extensor tibiae muscle fibre by long depolarizing pulses applied through an intracellular electrode. Increasing intensities of stimulation (a–i). (a): oscillatory "local responses" became spike-like but graded activities with increasing stimulus intensity. (b–h): frequency of pulses also increased. (i–l) and (m–p): repetitive stimulation of this fibre albeit at different intensities led progressively to a decrease in the capacity of the fibre to respond repetitively. (From Cerf *et al.,* 1959.)

The suggestion that the graded electrically-excited response of insect muscle fibres is due to an early rise in P_K resulting in incomplete expression of the depolarizing sodium (?) activation process receives some support from studies of the effects of alkaline earth ions or othopteran skeletal muscle fibres (Werman *et al.,* 1961). Alkaline earth ions such as barium, strontium and, to a lesser extent, calcium inactivate potassium. As a result, electrical stimulation of, for example, the barium treated muscle fibre leads to complete expression of the depolarizing electrogenesis with the appearance of all-or-none, rather than graded, electrically-excited responses. These action potentials propagate without decrement. Similar changes are produced by low concentrations (ca 10^{-6} M) of the alkaloid ryanodine (Usherwood, 1962b). The fact that both ryanodine and alkaline earth ions (Fig. 18A) increase the effective resistance (R_{eff}) and abolish the delayed rectification of orthopteran skeletal muscle

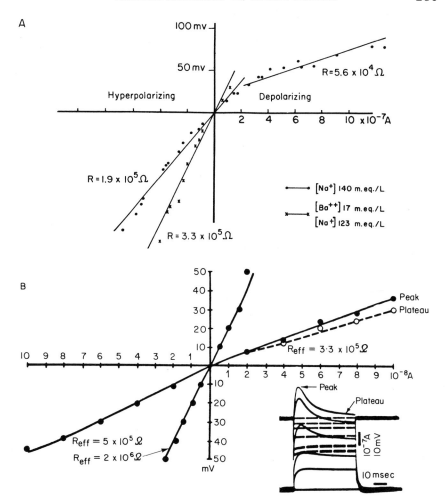

Fig. 18 A. Increased effective resistance of grasshopper extensor tibiae muscle fibres after treatment with saline in which 17 mEq/litre barium was substituted for an equal quantity of sodium. In the presence of this low concentration of barium the delayed rectification characteristic of the fibre in normal saline was abolished (not shown). The effective resistances derived from the slopes of the current-voltage curves are indicated. R is the effective resistance. (From Werman *et al.,* 1961.)

B. Current-voltage relations of two fibres of the rectal muscles of the cockroach, *Periplaneta americana.* The broken line represents the delayed rectification encountered in a fibre with a low effective resistance. Inset: Responses of a low resistance fibre to a wide range of depolarizing current pulses applied through an intracellular microelectrode. Note delayed rectification and small hyperpolarization following termination of the current pulse (potassium activation?). The upper trace (broken) represents current and the lower trace (solid) is voltage. (From Belton and Brown, 1969.)

fibres supports the contention that they inactivate potassium since both R_{eff} and delayed rectification are linked to P_K (Werman *et al.,* 1961). Similarly the all-or-none action potentials recorded from muscle fibres of *Romalea microptera* in the presence of barium and strontium are markedly prolonged, probably due to the low permeability of the muscle fibres to potassium which delays repolarization. However, there is some evidence that prolongation of the spike is also due partly to a decrease of sodium (?) inactivation. There is obviously a great need at this stage to study the roles of the major cations and anions present in insect muscle systems in the generation of graded and all-or-none electrically-excited responses of orthopteran and other insect muscle fibres using the voltage clamp technique. Not only would information be obtained on the contribution made by various ions to the depolarizing and repolarizing electrogenesis of the electrically-excited response, but also information on the time courses of the inward and outward currents during this response could be obtained using this technique. With this information we might then be in a position to account for the graded electrically-excited responsiveness of insect muscle fibres.

All that can be said at present about the electrically-excited response of insect muscle is that it is accompanied by a change in membrane permeability. This has been demonstrated by Werman *et al.* (1961) albeit using the all-or-none response of barium treated leg muscle fibres of *Romalea microptera.* They measured R_{eff} of these muscle fibres by applying brief hyperpolarizing current pulses and found, not surprisingly, that R_{eff} was much lower during the spike than at rest (Fig. 19).

The electrically-excited responses of a large number of different insect muscles from many different insect orders have now been studied and there is as yet no evidence that all-or-none action potentials normally occur in insect muscle. McCann and Boettiger (1961) obtained intracellular potentials from flight muscles of *Sarcophaga bullata,* the flesh fly, and *Vespula diabolique,* a wasp, which were termed all-or-none action potentials (Fig. 20). However, other data presented by McCann and Boettiger indicate the graded nature of these electrically-excited responses. Ikeda and Boettiger (1965b) are reported to have recorded all-or-none spikes from the fibrillar flight muscle fibres of the bee, *Bombus pennsylvanicus* (Hymenoptera). However, since the responses recorded from different points along the same fibre were somewhat variable, it is probable that these responses were also graded electrically-excited potentials.

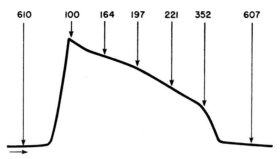

Fig. 19. Average relative impedances during all-or-none action potential of an insect muscle treated with barium. Data obtained from fourteen fibres of the extensor tibiae muscle of the grasshopper, *Romalea microptera*. This muscle normally responds to depolarization with a graded electrically-excited response but in the presence of barium ions all-or-none spikes are evoked during depolarization. (From Werman *et al.*, 1961.)

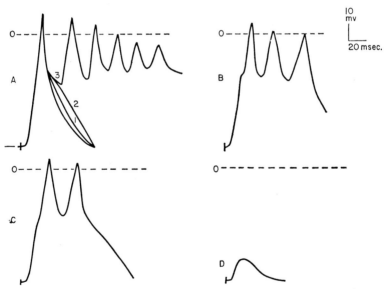

Fig. 20. Variations in responsiveness of beetle flight muscle fibres. A stimulus of constant intensity and duration given to the innervating nerve at 1 sec intervals evokes a series of muscle electrical responses which vary from a single large EPSP plus graded spike-like responses through a series of oscillatory responses to a small EPSP without an electrically-excited response. Following this last small response the cycle of events is repeated.

The records indicate considerable variation in the height of the EPSP (polyneuronal innervation?). The prolongation of the electrical response together with the oscillatory activity could be due to either prolongation of the EPSP or to inactivation of the repolarizing electrogenesis which terminates the electrically-excited depolarization. (Records from McCann and Boettiger, 1961.)

The electrically-excited responses of insect heart muscle are, at first sight, all-or-none in character. In the cockroach they consist of a slow depolarizing pre-potential followed by rapid depolarizing and repolarizing phases respectively (T. Miller, personal communication). The largest potentials never exceed 50 mV and never overshoot the zero membrane potential. Until these potentials have been studied in more detail it is difficult to decide whether they are all-or-none. However, the fact that they show considerable variation in magnitude suggests that this is unlikely. The same arguments probably apply equally to the electrically-excited responses of moth heart muscle fibres (McCann, 1965). These responses are, however, interesting in another respect since despite the apparent structural homogeneity that prevails throughout the heart, electrically-excited responses of different shapes can be recorded from different parts of the heart.

The other types of electrically-excited responses seen in insect muscle fibres are delayed rectification and anomalous rectification (Fig. 21). Delayed rectification has been demonstrated in a number of different muscle preparations (Cerf et al., 1959; Werman et al., 1961; Belton, 1969; Belton and Brown, 1969) (Fig. 18B) and probably results from potassium activation. Anomalous rectification has only recently been demonstrated in insect muscle in the retractor unguis muscle of the locust, *Schistocerca gregaria,* and could involve chloride and/or potassium conductance changes (Figs 6 and 21).

A. THE IONIC BASIS FOR DEPOLARIZING AND REPOLARIZING ELECTRICALLY-EXCITED ELECTROGENESIS

It has already been noted that due to the somewhat unusual nature of the ionic environment of some insect muscles the membranes of these muscles have evolved ionic properties which are often remarkably different from those of classical vertebrate bio-electric systems. However, it is still possible to describe the electrogenic activity of most insect muscle tissues on the basis of "equivalent circuit" theory, although the familiar roles of sodium and potassium ions in depolarizing and repolarizing electrically-excited electrogenesis may be assumed by other ions (Fig. 7). Even in the lower insects where the ionic properties of the haemolymph resemble those of vertebrate plasma, activity of synaptic and electrically excitable membranes may involve more ions than just sodium and potassium.

Wood (1963) measured the intracellular sodium content of skeletal muscle fibres of *Locusta migratoria, Periplaneta americana,* and *Carausius morosus.* The measurements were made after exposing the muscles to either haemolymph or salines containing different concentrations of sodium ions. He found that under all conditions the peak of the response to neural stimulation (i.e. EPSP plus

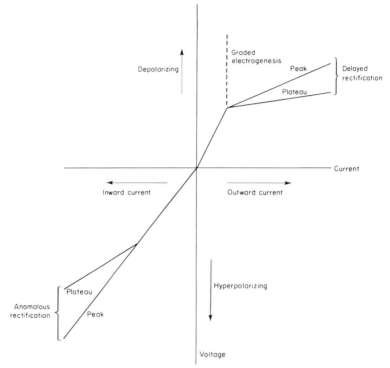

Fig. 21. Diagrammatic representation of current-voltage characteristics of locust skeletal muscle fibres.

electrically-excited response) in the locust and cockroach did not reach E_{Na} and as a result of this he suggested that sodium ions might be relatively impermeant during this type of activity. Wood correctly pointed out, however, that it is necessary to remember that the neurally-evoked response is a compound potential consisting of an EPSP plus an electrically-excited response. Furthermore, the magnitude of the electrically-excited response is dependent mainly on the magnitude of the EPSP since it is a graded event, not an all-or-none event. Therefore, even if the depolarizing component of the electrically-excited response is due to sodium activation, there is no reason to expect that the peak of this graded response will be at the equilibrium potential for sodium. It would be of value to repeat these studies whilst incorporating barium, strontium or ryanodine into the saline so that the effects of different concentrations of extracellular sodium on the all-or-none spike can be determined. At the same time the effects of tetrodotoxin, which specifically inactivates sodium, might be examined. The suggestion made by Werman *et al.* (1961) that the graded electrically-excited responses of insect muscle fibres could be due to an early rise in P_K is of interest here since this could explain why the peak of the electrically-excited response does not coincide with E_{Na}. Alternatively, ions other than sodium may carry the current during the electrically-excited response and these ions may have different equilibrium potentials than sodium. The fact that the peak response in zero sodium saline overshoots the predicted value for E_{Na} (Wood, 1963) suggests that this alternative proposal may well be true. One main difficulty, however, lies in accepting completely the values for the intracellular and extracellular sodium concentrations and therefore the predicted values for E_{Na} described by Wood (1963). The muscles he used for his studies were large and complex in structure and, therefore, not particularly suitable for studies involving changes in ionic environment. It would undoubtedly be profitable to repeat these experiments using either the much simpler retractor unguis muscle preparation (Usherwood, 1967b; Usherwood and Machili, 1968) or better still, a single muscle fibre preparation.

The possibility that ions other than sodium are involved in carrying some of the inward current during the depolarizing component of the electrically-excited response cannot of course be ignored, even in those insects where E_{Na} is normally markedly positive to the resting membrane potential. Nevertheless, the magnitude and rate of rise of the electrically-excited response of locust and stick insect skeletal muscle fibres are affected by variations in $[Na_o^+]$ suggesting perhaps that sodium ions carry at least part of the inward current in the skeletal muscles of these insects during electrically-excited activity.

In *Romalea microptera* the muscle fibres are surrounded by haemolymph containing relatively high concentrations of sodium (ca 100 mM). When the extensor tibiae muscle fibres of this insect are exposed to saline containing no sodium ions the graded electrically-excitable response disappears (Werman *et al.*, 1961). There is a suggestion here that sodium ions carry the inward current during depolarizing electrically-excited electrogenesis and this is supported by the fact that sodium is normally the major cation in the haemolymph of this insect. However, many other changes, apart from reduction of $[Na_o^+]$ to zero, must occur when, as in the experiments of Werman et al. (1961), sucrose is substituted for sodium (e.g. reduction in extracellular anion concentration; increase in resistance of extracellular or phase) and these could contribute either directly or indirectly to the loss of the electrically-excited response. When barium chloride is substituted for sodium chloride a depolarizing stimulus produces all-or-none spikes. If sodium is involved in generating the electrically-excited response then presumably barium ions can also carry the inward depolarizing current as well as inactivating potassium (Werman et al., 1961).

The sodium gradient across the membranes of skeletal (Huddart, 1966b) and heart (McCann, 1965) muscles of moths is reversed compared with that across, for example, the membranes of the quid giant axon and the locust skeletal muscle fibre. Huddart (1966a) found that the amplitude of the depolarizing component of the electrically-excited response of the leg muscle fibres of the moth, *Bombyx mori* (Lepidoptera), was only slightly affected by altering $[Na_o^+]$ although once again, since this is a graded response, it is difficult to evaluate the significance of amplitude changes. The response is not affected by replacement of sodium by choline or acetylcholine, two supposedly impermeant cations. Both tetrabutyl-ammonium and tetraethylammonium ions abolish the depolarizing electrically-excited response, although tetramethylammonium ions have no effect. Similar results have been obtained from the leg muscles of *Carausius morosus* (Wood, 1957). Normally, stimulation of the large excitatory axon to the flexor tibiae muscle of *Bombyx mori* evokes a large EPSP plus graded electrically-excited response (Huddart, 1966c). In most fibres of this muscle the combined response does not overshoot the zero potential level but in a few fibres with "high" resting potentials overshoots of 8 mV have been recorded. However, once again, the term overshoot is of little significance since the response contains a large synaptic component and it is the magnitude of this component which determines the height of the electrically-excitable component and also, in part, the total magnitude of the response of the muscle fibre to neural stimulation.

McCann has shown that moth heart muscle is still capable of electrical activity in the absence of extracellular sodium ions and similar observations have been made on heart muscle of the cockroach (T. Miller, personal communication). When extracellular sodium is removed from the locomotory muscles of the cabbage white butterfly and replaced by an osmotically equivalent amount of sucrose, the electrically-excited depolarizing electrogenesis is initially enhanced and then depressed (Belton, personal communication). If removal of extracellular sodium ions transiently enhances the depolarizing component of the electrically-excited response of these muscles, then it would seem reasonable to assume that sodium ions do not, in this case, carry the current during depolarization. In the stick insect an increase in external calcium concentration $[Ca_o^{++}]$ from 0 mM to 10 mM is accompanied by an increase of 36 mV in the magnitude of the response to neural stimulation (Wood, 1957). However, the fact that the neurally evoked response contains a large EPSP and that the electrically-excited component increases by only 3 mV between 0 mM and 10 mM $[Ca_o^{++}]$ suggests that the major effect of this ion is on the synaptic rather than the non-synaptic membrane. The contribution of magnesium to the total inward current during the depolarizing component of the electrically-excited response of *Carausius* muscle fibres seems to be much more significant. Wood (1957) found that the magnitude of this response was maximal when the external concentration of magnesium $[Mg_o^{++}]$ was between 50 mM and 70 mM. However, it should be emphasized once again that increases in magnitude of the electrically-excited response do not necessarily indicate changes in the magnitude of the inward current. They could equally well be explained on the basis of changes in the relative time courses of the repolarizing and depolarizing conductance changes. Once again, only careful studies using voltage clamp techniques will give us a clear-cut answer to this problem.

Apart from muscles of the stick insect, there is also some evidence that magnesium ions are involved in electrically-excited electrogenesis of other insect muscles. For example, Weevers (1966) suggests that this divalent cation may contribute to the inward depolarizing current during the electrically-excited response of caterpillar *(Antherea pernyi* and *myletta)* muscle fibres, since electrically-excited responses still occur in zero sodium saline, but only provided at least 20 mM magnesium is present. Also, when magnesium is increased beyond 20 mM the response height is increased. McCann (1967) has suggested that magnesium may also be involved in the production of the depolarizing component of the electrically-excited response of heart muscle fibres of *Hyalophora cecropia.* Here the magnesium equilibrium potential is about +13 mV. However, in this muscle it is rather interesting to note that when all magnesium was removed from the saline, depolarization of the muscle fibres resulted. This rather disturbing observation should certainly be investigated further. Possibly removal of the divalent cation affects the permeability of the muscle fibre, to one or more monovalent ions. The possibility that anions may play a role in the excitatory process has also been examined but since changes in electrically-excited activity, resulting from changes in external anion concentrations, are accompanied at times by marked changes in resting potential, it is difficult to evaluate their significance.

In the muscle fibres of the mealworm, *Tenebrio molitor,* electrically-excited depolarizing electrogenesis still occurs in the absence of extracellular sodium. Belton and Grundfest (1961) suggested that here also magnesium ions probably carry much of the inward positive current during the response.

Repolarizing electrogenesis seems to be brought about in most insect muscles, as in the squid giant axon and probably in frog skeletal muscles as well, by an increase in P_K, although the role, if any, of anions in repolarizing electrogenesis has not been adequately studied as yet. Tetraethylammonium ions in concentrations less than 1 mM convert the graded electrically-excited responses of *Romalea microptera* skeletal muscle fibres to all-or-none responses and prolonged oscillatory responses occur during depolarization with applied current. These observations are consistent with inactivation of potassium, as is the increase (ca $\times$ 2) in R_{eff}. Since the resting membrane in this insect is a reasonably good potassium electrode, this fall in P_K is not sufficient to significantly alter the resting potential. Procaine (1 mM) increases R_{eff} of *Romalea* skeletal muscle fibres and also *Tenebrio* skeletal muscle fibres and converts the graded electrically-excited responses of these fibres to all-or-none spikes. The effect of procaine on the graded response of *Tenebrio* muscle fibres is easily explained if it inactivates potassium, since repolarizing electrogenesis in these muscles is due to potassium activation. However, the effect of procaine on R_{eff} of mealworm muscle fibres is more difficult to explain, in view of the poor potassium-electrode properties of the resting muscle fibres of this insect, unless procaine inactivates other ions as well as potassium. Barium, strontium and choline chloride also convert the graded electrically-excited responses of *Tenebrio* muscle fibres to all-or-none responses (Belton and Grundfest, 1961).

The electrically-excited responses of *Tenebrio* muscle fibres recorded at low levels of $[K_o^+]$ have a terminal undershoot which can be about as large as the initial depolarizing phase (Belton and Grundfest, 1962). When $[K_o^+]$ is increased both depolarizing and repolarizing phases become smaller. Belton and Grundfest (1962) found that when *Tenebrio* muscle fibres are exposed to saline containing 200 mEq/L potassium the resting potential does not change. Stimulation of the muscle fibres in this condition, with brief suprathreshold depolarizing currents, leads to the appearance of spike-like responses with long lasting plateaus of depolarization. R_{eff} of the muscle fibres decreases relative to the resting value during the plateau. It is assumed that the long lasting depolarizations result from a process of potassium activation which has a slow onset. Barium ions, which normally convert the graded electrically-excited response of insect muscle fibres to an all-or-none response, decrease the duration of the depolarizing response of mealworm muscle fibres in high potassium saline but in low potassium saline containing barium ions the graded electrically-excited responses are prolonged. Once again the action of barium is explicable in terms of potassium inactivation.

Belton and Grundfest's (1961, 1962) studies clearly demonstrate that the membrane of mealworm skeletal muscle fibres becomes a potassium electrode during electrically-excited activity. When $[K_o^+]$ is high the depolarizing response resulting from a suprathreshold electrical stimulus is spike-like. However, the possibility that even in this high potassium environment some component of this electrically-excited response is caused by an increase in permeability of the membrane to ions other than potassium cannot be ignored. It is perhaps unwise therefore to call this response a potassium-spike (Belton and Grundfest, 1961, 1962). In a medium containing a high concentration of potassium E_K is probably well above (more positive than) the equilibrium value for the initial depolarizing electrogenesis, so that the delayed increase in P_K which normally (in a low potassium environment) repolarizes the membrane now produces an added depolarization. However, even in high potassium saline E_K and the value of the membrane potential at the peak of the electrically-excited depolarization do not coinicide,* which suggests that other ions as well as potassium are involved in generating the spike-like response.

---

* This is probably not due to graded responsiveness of the electrically excitable membrane.

The effects of metabolic inhibitors on the electrically-excited response of insect muscle fibres have received only scant attention. McCann (1967) found that DNP caused a gradual diminution of amplitude, a progressive suppression of the plateau phase of the response, an exaggeration of the slow pacemaker-type prepotential and finally the appearance of a biphasic abortive response.

The non-synaptic membrane of some insect muscle fibres either produces only small depolarizing electrically-excitable responses, as in the locust spiracular muscle (Hoyle, 1959), or is completely electrically inexcitable as in the tonic fibres of the locust meta-thoracic extensor tibiae muscle (Cerf *et al.*, 1961; Usherwood and Grundfest, 1965; Usherwood, 1967b; Cochrane *et al.*, 1968). In the visceral (rectal) muscle fibres of the cockroach (Belton and Brown, 1969) there is little evidence for depolarizing electrogenesis, although the fibres do show delayed rectification (potassium activation?) (Fig. 18B).

ACKNOWLEDGEMENTS

The author is indebted to Professor H. Grundfest for permission to quote from a number of his pending publications and to Dr N. C. Spurway and Dr J. W. P. Barnes for much helpful criticism and discussion during the preparation of this manuscript.

REFERENCES

Adrian, R. H. (1958). The effects of membrane potential and external potassium concentration on the potassium permeability of muscle fibres. *J. Physiol. Lond.* **143**, 59–60P.

Adrian, R. H. (1960). Potassium chloride movement and the membrane potential of frog muscle. *J. Physiol. Lond.* **151**, 154–185.

Aidley, D. J. (1967). The excitation of insect skeletal muscles. In *Advances in Insect Physiology* (J. W. L. Beament, J. E. Treherne and V. B. Wigglesworth, eds). Vol. 4, pp. 1–31. Academic Press, London and New York.

Arvy, L. (1952). Particularités histologiques des centres leucopoiétiques thoraciques ches quelques Lépidoptères. *C.r. hebd. Séanc. Acad. Sci. Paris.* **235**, 1539–1541.

Ashhurst, D. E. (1967). The fibrillar flight muscles of giant water-bugs: an electron microscope study. *J. Cell. Sci.* **2**, 435–444.

Ashhurst, D. E. (1968). The connective tissue of insects. *A. Rev. Ent.* **13**, 45–74.

Belton, P. (1958). Membrane potentials recorded from moth muscle fibres. *J. Physiol. Lond.* **142**, 20–21P.

Belton, P. (1969). Innervation and neural excitation of ventral muscle fibres of the larva of the wax moth, *Galleria mellonella* L. *J. Insect Physiol.* **15**, 731–741.

Belton, P. and Brown, B. E. (1969). The electrical activity of cockroach visceral muscle fibers. *Comp. Biochem. physiol.* **28**, 853–863.

Belton, P. and Grundfest, H. (1961). Comparative effects of drugs on graded responses of insect muscle fibers. *Fedn Proc. Fedn Am. Socs exp. Biol.* **20**, 339.

Belton, P. and Grundfest, H. (1962). Potassium activation and K spikes in muscle fibers of the mealworm larva *(Tenebrio molitor).* *Am. J. Physiol.* **203**, 588–594.

Bennett, M. V. L. (1964). Nervous function at the cellular level. *A. Rev. Physiol.* **26**, 289–340.

Bennett, M. V. L., Aljure, E., Nakajima, T. and Pappas, G. D. (1963). Electronic junctions between teleost spinal neurons. Electrophysiology and ultrastructure. *Science, N.Y.* **141**, 262–264.

Beránek, R. and Miller, P. (1968a). Sensitivity of insect muscle fibres to L-glutamate. *J. Physiol. Lond.* **196**, 71–72P.

Beránek, R. and Miller, P. (1968b). The action of iontophoretically applied glutamate on insect muscle fibres. *J. exp. Biol.* **49**, 83–93.

Bernstein, J. (1902). Untersuchungen zur Thermodynamik der bioelektrischen Ströme. *Pflügers Arch. ges. Physiol.* **92**, 521–562.

Bernstein, J. (1912). "Electrobiologie". Braunschweig: Fr. Vieweg.

Bishop, G. H., Briggs, A. P. and Ronzoni, E. (1925). Body fluid of the honeybee larva: II. Chemical constituents of the blood and their osmotic effects. *J. biol. Chem.* **66**, 77–88.

Boistel, J. and Fatt, P. (1958). Membrane permeability change during inhibitory transmitter action in crustacean muscle. *J. Physiol. Lond.* **144**, 176–191.

Boyle, P. J. and Conway, E. J. (1941). Potassium accumulation in muscle and associated changes. *J. Physiol. Lond.* **100**, 1–63.

Brady, J. (1967a). Haemocytes and the measurement of potassium in insect blood. *Nature, Lond.* **215**, 96–97.

Brady, J. (1967b). The relationship between blood ions and blood-cell density in insects. *J. exp. Biol.* **47**, 313–326.

Brady, J. (1968). Control of the circadian rhythm of activity in the cockroach III. A possible role of the blood-electrolytes. *J. exp. Biol.* **49**, 39–47.

Brandt, P. W., Reuben, J. P., Girardier, L. and Grundfest, H. (1965). Correlated morphological and physiological studies on isolated single muscle fibres. 1. Fine structure of the crayfish muscle fibre. *J. Cell Biol.* **25**, 233–261.

Brown, F. and Danielli, J. F. (1964). The cell surface and cell physiology. In *Cytology and Cell Physiology.* (G. H. Bourne, ed.). pp. 239–310. Academic Press, London and New York.

Buck, J. B. (1953). In *Insect Physiology* (K. D. Roeder, ed.). Chapter 6. Wiley, New York.

Carrington, C. A. and Tenney, S. M. (1959). Chemical constituents of haemolymph and tissue in *Telea polyphemus* Cram. with particular reference to the question of ion binding. *J. Insect Physiol.* **3**, 402–413.

Cerf, J. A., Grundfest, H., Hoyle, G. and McCann, F. V. (1959). The mechanism of dual responsiveness in muscle fibers of the grasshopper *Romalea microptera. J. gen. Physiol.* **43**, 377–395.

Clark, E. W. (1958). A review of the literature on calcium and magnesium in insects. *Ann. ent. Soc. Am.* **51**, 142–154.

Clark, E. W. and Craig, R. (1953). The calcium and magnesium content in the haemolymph of certain insects. *Physiol. Zöol.* **26**, 101–107.

Cochrane, D. G. and Elder, H. Y. (1967). Morphological changes in insect muscles during influx and efflux of potassium ions, chloride ions and water. *J. Physiol. Lond.* **191**, 30–31P.

Cochrane, D. G., Elder, H. Y. and Usherwood, P. N. R. (1969). Electrical, mechanical and ultrastructural properties of tonic and phasic muscle fibres in the locust *(Schistocerca gregaria). J. Physiol. Lond.* **20**, 60–69P.

Davey, K. G. (1964). The control of visceral muscles in insects. In *Advances in Insect Physiology* (J. W. L. Beament, J. E. Treherne and V. B. Wigglesworth, eds). Vol. 2, pp. 219–245. Academic Press, London and New York.

del Castillo, J., Hoyle, G. and Machne, X. (1953). Neuromuscular transmission in a locust. *J. Physiol. Lond.* **121**, 539–547.

Donnan, F. G. (1924). The theory of membrane equilibrium. *Chemy Ind. Rev.* **1**, 73–90.

Duchâteau, G. and Florkin, M. (1958). A survey of aminoacidaemias with special references to the high concentrations of free amino acids in insect haemolymph. *Archs int. Physiol. Biochim.* **66**, 573–591.

Duchâteau, G., Florkin, M. and Leclercq, J. (1953). Concentrations des bases fixes et des types de compositions de la base totale de l'hémolymph des insectes. *Archs int. Physiol. Biochim.* **61**, 518–549.

Eccles, J. C. (1964). *The Physiology of Synapses.* 316 pp. Springer-Verlag, Berlin, Gottingen, Heidelberg.

Edwards, G. A. and Challice, C. E. (1960). The ultrastructure of the heart of the cockroach, *Blattella germanica. Ann. ent. Soc. Am.* **53**, 369–383.

Edwards, G. A., Ruska, H. and de Harven, E. (1958). Electron microscopy of peripheral nerves and neuromuscular junctions in the wasp leg. *J. biophys. biochem. Cytol.* **4**, 107–114.

Fatt, P. and Katz, B. (1951). An analysis of the end-plate potential recorded with an intracellular electrode. *J. Physiol. Lond.* **115**, 320–370.

Fawcett, D. W. (1966). An atlas of fine structure. *The Cell: Its Organelles and Inclusions.* 448 pp. W. B. Saunders Company, Philadelphia.

Florkin, M. and Jeuniaux, Ch. (1964). Haemolymph: Composition. In *The Physiology of the Insecta* (M. Rockstein, ed.). Vol. III, pp. 109–152. Academic Press, London and New York.

Foulkes, J. G., Pacey, J. A. and Perry, F. A. (1965). Contractures and swelling of the transverse tubules during chloride withdrawal in frog skeletal muscle. *J. Physiol. Lond.* **180**, 96–115.

Goldman, D. E. (1943). Potential, impedance and rectification in membranes. *J. gen. Physiol.* **27**, 37–60.

Grundfest, H. (1956). Some properties of excitable tissues. *Trans. 5th Josiah Macy Jr. Conf. on Nerve Impulses,* pp. 177–218.

Grundfest, H. (1957). Electrical inexcitability of synapses and some of its consequences in the central nervous system. *Physiol. Rev.* **37**, 337–361.

Grundfest, H. (1959). Synaptic and ephaptic transmission. In *Handbook of Physiology. Neurophysiology.* Washington, D.C. *Am. Physiol. Soc.* **1**, 147–197.

Grundfest, H. (1964). Bioelectrogenesis. In *Encyclopaedia of Electrochemistry* (C. A. Hampel, ed.). pp. 107–113. Reinhold, New York.

Grundfest, H. (1966). Comparative electrobiology of excitable membranes. *Adv. comp. Physiol. Biochem.* **2**, 1–116.

Guthrie, D. M. (1962). Control of the ventral diaphragm in an insect. *Nature, Lond.* **196**, 1010–1012.

Hagiwara, S. and Watanabe, A. (1954). Action potential of insect muscle re-examined with intra-cellular microelectrode. *Jap. J. Physiol.* **4**, 65–78.

Hama, K. (1961). Some observations on the fine structure of the giant fibres of the crayfish *(Cambarus virilis* and *Cambaras clarkii)*, with special reference to the sub-microscopic organisation at the synapses. *Anat. Rec.* **141**, 275–280.

Hamori, J. (1961). Cholinesterase in insect muscle innervation with special reference to insecticide effects of DDT and DFP. *Bibliotheca Anat.* **2**, 194–206.

Hanai, T., Haydon, D. A. and Taylor, J. (1965). Some further experiments in bimolecular lipid membranes. *J. gen. Physiol.* **48**, 59–63.

Harlow, P. A. (1958). The action of drugs on the nervous system of the locust *(Locusta migratoria)*. *Ann. appl. Biol.* **46**, 55–73.

Hays, E. A., Lang, M. A. and Gainer, H. (1968). A re-examination of the Donnan distribution as a mechanism for membrane potentials and potassium and chloride ion distributions in crab muscle fibres. *Comp. Biochem. Physiol.* **26**, 761–792.

Hill, R. B. and Usherwood, P. N. R. (1961). The action of 5-hydroxytryptamine and related compounds on neuromuscular transmission in the locust *Schistocerca gregaria*. *J. Physiol. Lond.* **157**, 393–401.

Hodgkin, A. L. and Horowicz, P. (1957). The influence of potassium and chloride ions on the membrane potential of single muscle fibres. *J. Physiol. Lond.* **148**, 127–160.

Hodgkin, A. L. and Huxley, A. F. (1952). A quantitative description of membrane current and its application to conduction and excitation in nerve. *J. Physiol. Lond.* **117**, 500–544.

Hodgkin, A. L. and Katz, B. (1949). The effect of sodium ions on the electrical activity of the giant axon of the squid. *J. Physiol. Lond.* **108**, 37–77.

Hodgkin, A. L. and Keynes, R. D. (1955). Active transport of cations in giant axons from *Sepia* and *Loligo*. *J. Physiol. Lond.* **128**, 28–60.

Hodgkin, A. L. and Rushton, W. A. H. (1946). The electrical constants of a crustacean nerve fibre. *Proc. R. Soc.* B **133**, 444–479.

Hoyle, G. (1952). "Slow" and "fast" nerve fibres in locusts. *Nature, Lond.* **172**, 165.

Hoyle, G. (1954). Changes in the blood potassium concentration of the African migratory locust *(Locusta migratoria migratoriodes* R. and F.) during food deprivation, and the effect on neuromuscular activity. *J. exp. Biol.* **31**, 260–270.

Hoyle, G. (1955). The effects of some common cations on neuromuscular transmission in insects. *J. Physiol. Lond.* **127**, 90–103.

Hoyle, G. (1957). Nervous control of insect muscles. In *Recent Advances in Invertebrate Physiology* (B. T. Scheer, ed.). pp. 73–98. Eugene University of Oregon Press.

Hoyle, G. (1959). The neuromuscular mechanisms of an insect spiracular muscle. *J. Insect Physiol.* **3**, 378–394.

Hoyle, G. (1962). Comparative physiology of conduction in nerve and muscle. *Amer. Zool.* **2**, 5–25.

Hoyle, G. (1965). Nature of the excitatory sarcoplasmic reticular junction. *Science, N.Y.* **149**, 70–72.

Hoyle, G. (1966a). An isolated insect ganglion-nerve-muscle preparation. *J. exp. Biol.* **44**, 413–427.

Hoyle, G. (1966b). Functioning of the inhibitory-conditioning axon innervating insect muscles. *J. exp. Biol.* **44**, 429–435.

Hubbard, J. I. and Løyning, Y. (1966). The effect of hypoxia on neuromuscular transmission in a mammalian preparation. *J. Physiol. Lond.* **185**, 205–223.

Huddart, H. (1966a). The effect of sodium ions on resting and action potentials in skeletal muscle fibre of *Bombyx mori* (L.). *Arch. int. Physiol. Biochim.* **74**, 592–602.

Huddart, H. (1966b). Ionic composition of haemolymph and myoplasm in Lepidoptera in relation to their membrane potentials. *Arch. int. Physiol. Biochim.* **74**, 603–613.

Huddart, H. (1966c). Electrical and mechanical responses recorded from Lepidopterous skeletal muscle. *J. Insect Physiol.* **12**, 537–545.

Huddart, H. (1966d). The effect of potassium ions on resting and action potentials in lepidopterous muscle. *Comp. Biochem. Physiol.* **18**, 131–140.

Huddart, H. (1967). The effect of chloride ions on moth skeletal muscle fibres. *Comp. Biochem. Physiol.* **20**, 355–361.

Huddart, H. and Wood, D. W. (1966). The effect of DNP on the resting potential and ionic content of some insect skeletal muscle fibres. *Comp. Biochem. Physiol.* **18**, 681–688.

Ikeda, K. and Boettiger, E. G. (1965a). Studies on the flight mechanism of insects. II. The innervation and electrical activity of the fibrillar muscles of the bumble bee, *Bombus. J. Insect Physiol.* **11**, 779–789.

Ikeda, K. and Boettiger, E. G. (1965b). Studies on the flight mechanism of insects. III. The innervation and electrical activity of the basalar fibrillar flight muscle of the beetle, *Oryctes rhinoceros. J. Insect Physiol.* **11**, 791–802.

Iyotami, K. and Kaneshina, K. (1958). Localization of cholinesterase in the American cockroach. *Jap. J. appl. Ent. Zool.* **2**, 1–10.

Jones, J. C. (1964). The circulatory system of insects. In *Physiology of Insecta* (M. Rockstein, ed.). Vol. III, pp. 1–107. Academic Press, London and New York.

Katz, B. (1962). The Croonian Lecture: The transmission of impulses from nerve to muscle, and the subcellular unit of synaptic action. *Proc. R. Soc.* B **155**, 455–477.

Katz, B. (1966). *Nerve, Muscle and Synapse.* 193 pp. McGraw-Hill, New York and London.

Kerkut, G. A., Shapira, A. and Walker, R. J. (1965). The effect of acetylcholine, glutamic acid and GABA on the contractions of the perfused cockroach leg. *Comp. Biochem. Physiol.* **16**, 37–48.

Kerkut, G. A. and Walker, R. J. (1966). The effect of L-glutamate, acetylcholine and gamma-aminobutyric acid on the miniature end-plate potentials and contractures of the coxal muscles of the cockroach, *Periplaneta americana. Comp. Biochem. Physiol.* **17**, 435–454.

Kernan, R. P. (1967). Electrogenic potassium pump related to generation of end-plate potentials in muscle. *Nature, Lond.* **214**, 725–726.

Keynes, R. D. (1954). The ionic fluxes in frog muscle. *Proc. R. Soc.* B **142**, 359–382.

Kusano, K. and Grundfest, H. (1967). Ionic requirements for synaptic electrogenesis in neuromuscular transmission of mealworm larvae *(Tenebrio molitor)*. *J. gen. Physiol.* **50**, 1092.

Langley, O. K. and Landon, D. N. (1968). A light and electron histochemical approach to the node of Ranvier and myelin of peripheral nerve fibers. *J. Histochem. Cytochem.* **15**, 722–731.

Levenbook, L. (1950). The composition of the horse bot fly *(Gastrophilus intestinalis)* larva blood. *Biochem. J.* **47**, 336–346.

Ludwig, D. (1951). Composition of the blood of the Japanese beetle *(Popillia japonica)*. *Physiol. Zool.* **24**, 329–334.

Malpus, C. M. (1968). Electrical responses of muscle fibres of dragonfly larvae in relation to those of other insects and of crustaceans. *J. Insect Physiol.* **14**, 1285–1301.

McCann, F. V. (1963). Electrophysiology of an insect heart. *J. gen. Physiol.* **46**, 803–821.

McCann, F. V. (1964). The effect of anion substitution on bioelectric potentials in the moth heart. *Comp. Biochem. Physiol.* **13**, 179–188.

McCann, F. V. (1965). Unique properties of the moth myocardium. *Ann. N.Y. Acad. Sci.* **127**, 84–99.

McCann, F. V. (1967). The effects of metabolic inhibitors in the moth heart. *Comp. Biochem. Physiol.* **20**, 411–430.

McCann, F. V. and Boettiger, E. G. (1961). Studies of the flight mechanism of insects. 1. The electrophysiology of fibrillar flight muscle. *J. gen. Physiol.* **45**, 125–142.

McCann, F. V. and Reece, R. W. (1967). Neuromuscular transmission in insects: effect of injected chemical agents. *Comp. Biochem. Physiol.* **21**, 115–124.

Morrison, G. D. (1928). The muscles of the adult honey-bee *(Apis mellifera* L.). *Q. Jl. microsc. Sci.* **71**, 395–463.

Peachey, L. D. and Huxley, A. F. (1964). Transverse tubules in crab muscle. *J. Cell. Biol.* **23** (70A).

Pringle, J. W. S. (1957). *Insect Flight.* 132 pp. Cambridge University Press.

Rambourg, A. and Leblond, C. P. (1967). Electron microscope observations on the carbohydrate-rich cell coat present at the surface of cells in the rat. *J. Cell. Biol.* **32**, 27–53.

Ramsay, J. A. (1955). The excretory system of the stick insect, *Dixippus morosus* (Orthoptera: Phasmidae). *J. exp. Biol.* **32**, 183–199.

Reuben, J. P., Girardier, L. and Grundfest, H. (1964). Water transfer and cell structure in isolated crayfish muscle fibers. *J. gen. Physiol.* **47**, 1141–1174.

Robertson, J. D. (1960). The molecular structure and contact relationship of cell membranes. *Prog. Biophys. biophys. Chem.* **10**, 343–418.

Sanger, J. W. and McCann, F. V. (1968). Ultrastructure of the myocardium of the moth, *Hyalophora cecropia. J. Insect Physiol.* **14**, 1105–1112.

Shanes, A. M. (1958). Electrochemical aspects of physiological and pharmacological action in excitable cells. 1. The resting cell and its alteration by extrinsic factors. *Pharmac. Rev.* **10**, 59–164.

Smith, D. S. (1961). The structure of insect fibrillar muscle. *J. Cell Biol.* **10**, 123–158.

Smith, D. S. (1965). The flight muscles of insects. *Scient. Am.* **212**, 76–88.

Smith, D. S. (1966). The organization and function of the sarcoplasmic reticulum and T-system of muscle cells. *Prog. Biophys. biophys. Chem.* **16**, 107–142.

Smith, D. S., Gupta, B. L. and Smith, U. (1966). The organization and myofilament array of insect visceral muscles. *J. Cell. Sci.* **1**, 49–57.

Stobbart, R. H. and Shaw, J. (1964). Salt and water: balance excretion. In *The Physiology of Insecta* (M. Rockstein, ed.). Vol. III, pp. 189–258. Academic Press, London and New York.

Sutcliffe, D. W. (1962). The composition of haemolymph in aquatic insects. *J. exp. Biol.* **39**, 325–343.

Sutcliffe, D. W. (1963). The chemical composition of haemolymph in insects and some other arthropods, in relation to their phylogeny. *Comp. Biochem. Physiol.* **9**, 121–135.

Takeuchi, A. and Takeuchi, N. (1960). On the permeability of the end-plate membrane during the action of the transmitter. *J. Physiol. Lond.* **154**, 52–67.

Tasaki, I. (1968). *Nerve excitation—a macromolecular approach.* 201 pp. C. C. Thomas. Springfield, Illinois, U.S.A.

Tasaki, I. and Singer, I. (1965). A macromolecular approach to the excitable membrane. *J. cell. comp. Physiol.* **66**, 137–146.

Tasaki, I. and Singer, I. (1968). Some problems involved in electrical measurements of biological systems. *Ann. N.Y. Acad. Sci.* **148**, 36–53.

Tauber, O. E. and Yeager, J. F. (1934). On total hemolymph (blood) cell counts in insects. *Ann. ent. Soc. Am.* **28**, 239–240.

Tauber, O. E. and Yeager, J. F. (1936). On the total hemolymph (blood) cell counts of insects. II. Neuroptera, Coleoptera, Lepidoptera and Hymenoptera. *Ann. ent. Soc. Am.* **29**, 112–118.

Tobias, J. M. (1948). Potassium, sodium and water interchange in irritable tissues and haemolymph of an omniverous insect, *Periplaneta americana. J. cell. comp. Physiol.* **31**, 125–142.

Treherne, J. E. (1961). Sodium and potassium fluxes in the abdominal nerve cord of the cockroach, *Periplaneta americana* L. *J. exp. Biol.* **38**, 315–322.

Usherwood, P. N. R. (1961). Spontaneous miniature potentials from insect muscle fibres. *Nature, Lond.* **191**, 814–815.

Usherwood, P. N. R. (1962a). The nature of the "slow" and "fast" contractions in the coxal muscles of the cockroach. *J. Insect Physiol.* **8**, 31–52.

Usherwood, P. N. R. (1962b). The action of the alkaloid ryanodine on insect skeletal muscle. *Comp. Biochem. Physiol.* **6**, 181–199.

Usherwood, P. N. R. (1963a). Spontaneous miniature potentials from insect muscle fibres. *J. Physiol. Lond.* **169**, 149–160.

Usherwood, P. N. R. (1963b). Response of insect muscle to denervation. I. Resting potential changes. *J. Insect Physiol.* **9**, 247–255.

Usherwood, P. N. R. (1963c). Response of insect muscle to denervation. II. Changes in neuromuscular transmission. *J. Insect Physiol.* **9**, 811–825.

Usherwood, P. N. R. (1967a). Permeability of insect muscle fibres to potassium and chloride ions. *J. Physiol. Lond.* **191**, 29–30P.

Usherwood, P. N. R. (1967b). Insect neuromuscular mechanisms. *Am. Zool.* **7**, 553–582.

Usherwood, P. N. R. (1968). A critical study of the evidence for peripheral inhibitory axons in insects. *J. exp. Biol.* **49**, 201–222.

Usherwood, P. N. R. (1969). Glutamate sensitivity of denervated insect muscle fibres. *Nature, Lond.* **223**, 411–413.

Usherwood, P. N. R., Cochrane, D. G. and Rees, D. (1968a). Changes in the structural, physiological and pharmacological properties of insect excitatory nerve-muscle synapses after motor nerve section. *Nature, Lond.* **218**, 589–591.

Usherwood, P. N. R. and Grundfest, H. (1964). Inhibitory post-synaptic potentials in grasshopper muscle. *Science, N.Y.* **143**, 817–818.

Usherwood, P. N. R. and Grundfest, H. (1965). Peripheral inhibition in skeletal muscle of insects. *J. Neurophysiol.* **28**, 497–518.

Usherwood, P. N. R. and Machili, P. (1966). Chemical transmission at the insect excitatory neuromuscular synapse. *Nature, Lond.* **210**, 634–636.

Usherwood, P. N. R. and Machili, P. (1968). Pharmacological properties of excitatory neuromuscular synapses in the locust. *J. exp. Biol.* **49**, 341–361.

Usherwood, P. N. R., Machili, P. and Leaf, G. (1968b). L-glutamate at insect excitatory nerve-muscle synapses. *Nature, Lond.* **219**, 1169–1172.

Van Asperen, K. and Van Esch, I. (1956). The chemical composition of the haemolymph in *Periplaneta americana. Archs néerl. Zool.* **11**, 342–360.

Van der Kloot, W. G. (1963). The electrophysiology and the nervous control of the spiracular muscle of pupae of the giant silkmoths. *Comp. Biochem. Physiol.* **9**, 317–334.

Weevers, R. de G. (1966). A lepidopteran saline: effects of inorganic cation concentrations on sensory, reflex and motor responses in a herbivorous insect. *J. exp. Biol.* **44**, 163–175.

Weidmann, S. (1952). The electrical constants of Purkinje fibres. *J. Physiol. Lond.* **118**, 348–360.

Werman, R., McCann, F. V. and Grundfest, H. (1961). Graded and all-or-none electrogenesis in arthropod muscle. 1. The effects of alkali-earth cations on the neuromuscular system of *Romalea microptera. J. gen. Physiol.* **44**, 979–995.

Wigglesworth, V. B. (1965). *The Principles of Insect Physiology.* 741 pp. Methuen & Co. Ltd., London. Dutton & Co. Inc., New York.

Wood, D. W. (1957). The effect of ions upon neuromuscular transmission in a herbivorous insect. *J. Physiol. Lond.* **138**, 119–139.

Wood, D. W. (1963). The sodium and potassium composition of some insect skeletal muscle fibres in relation to their membrane potentials. *Comp. Biochem. Physiol.* **9**, 151–159.

Wood, D. W. (1965). The relationships between chloride ions and resting potential in skeletal muscle fibres of the locust and cockroach. *Comp. Biochem. Physiol.* **15**, 303–312.

Wyatt, G. R. (1961). The biochemistry of insect haemolymph. *A. Rev. Ent.* **6**, 75–102.

Wyatt, G. R., Loughheed, T. C. and Wyatt, S. S. (1956). The chemistry of insect haemolymph. Organic components of the haemolymph of the silkworm, *Bombyx mori,* and two other species. *J. gen. Physiol.* **39**, 853–868.

Author Index

Numbers in italics are the pages on which the references are listed

Subject Index

Acetylcholine
  and electrically excitable membranes, 267
  and luminescence, 60–61, 74, 79
  and synaptic membranes, 243
Acheta (Gryllus) domesticus
  egg, dopa, 172
  nervous system, development, 100, 101, 105, 107, 116, 120, 121
  nervous system, regeneration, 126, 128, 129
Acrolita naevana, egg, frost resistance, 27
Actias selene
  ions in muscle systems, 220–221
  resting membrane potential, 239
Actin
  in skeletal muscle, 206
  in transverse tubular system, 210
Action potential, all-or-none, 257–259, 262
Adenosine-3',5'-phosphate, cyclic, and luminescence, 79
Adenyl cyclase, and luminescence, 79
Adrenaline and luminescence, 74, 75, 76, 77
Adult
  nervous system, development, 98, 106–107, 112–113, .115, 117, 118, 120, 121
  nervous system, regeneration, 125
Aedes, eye
  development, 111
  A. aegypti, A. mascariensis, pterines, 151, 158
Aeschna
  eye development, 112
  muscle fibre electrical constant, 212–213
Aeschna cyanea, pterines, 153
Aeschna grandis, haemolymph, 216–218
Afferent fibres
  development, 112, 119
  regeneration, 127, 128

Agrotis fucosa, larva, frost resistance, 28
Alcaligenes faecalis, pterine metabolism, 168
Alcohols, polyhydric, and frost resistance, 26–34
All-or-none spike, 257, 258, 259, 262, 264, 265, 266, 267, 269, 270
Alsophila pometaria, egg, frost resistance, 27, 28
Amathes ditrapezium, larva, frost resistance, 28
Amino acids
  decarboxylic, and synaptic membranes, 248
  in haemolymph, 215, 217, 218
  and membrane potential, 224, 232, 237
  p-aminobenzoic acid, folic acid synthesis, 185
  γ-aminobutyric acid (GABA), effect on inhibitory synaptic membranes, 252, 254–256
Ammonia, in extracellular fluid, 218
Ammonium salts, quaternary, effect on electrically excitable membranes, 267, 269
Amphetamine, and luminescence, 74, 75, 77, 79
Anagasta kühniella, larva, freezing, 19
Anatomy of firefly lantern, 54–59
Anguillula silusiae, frost resistance, 39
Animals other than insects
  amphibians, 140
  Anacystis, 180
  Callinectes, 236
  chick embryo, 124
  Crustacea, 127, 228, 236, 244
  frog, 223, 224, 225, 228, 231, 269
  Limulus, 100
  mammals, 17, 33, 34, 39, 105, 172, 216
  mouse, 124
  reptiles, 140

287

Cumulative List of Authors

Numbers in bold face indicate the volume number of the series

Aidley, D. J., **4**, 1
Andersen, Sven Olav, **2**, 1
Asahina, E., **6**, 1
Beament, J. W. L., **2**, 67
Boistel, J., **5**, 1
Burkhardt, Dietrich, **2**, 131
Bursell, E., **4**, 33
Burtt, E. T., **3**, 1
Carlson, A. D., **6**, 51
Catton, W. T., **3**, 1
Chen, P. S., **3**, 53
Colhoun, E. H., **1**, 1
Cottrell, C. B., **2**, 175
Dadd, R. H., **1**, 47
Davey, K. G., **2**, 219
Edwards, John S., **6**, 97
Gilbert, Lawrence I., **4**, 69
Harmsen, Rudolf, **6**, 139
Harvey, W. R., **3**, 133
Haskell, J. A., **3**, 133
Hinton, H. E., **5**, 65
Kilby, B. A., **1**, 111
Lees, A. D., **3**, 207
Miller, P. L., **3**, 279
Narahashi, Toshio, **1**, 175
Neville, A. C., **4**, 213
Pringle, J. W. S., **5**, 163
Rudall, K. M., **1**, 257
Shaw, J., **1**, 315
Smith, D. S., **1**, 401
Stobbart, R. H., **1**, 315
Treherne, J. E., **1**, 401
Usherwood, P. N. R., **6**, 205
Waldbauer, G. P., **5**, 229
Weis-Fogh, Torkel, **2**, 1
Wigglesworth, V. B., **2**, 247
Wilson, Donald M., **5**, 289
Wyatt, G. R., **4**, 287
Ziegler, Irmgard, **6**, 139

Cumulative List of Chapter Titles

Numbers in bold face indicate the volume number of the series